# Wireless Communications:
# The Future

# Wireless Communications: The Future

Professor William Webb

John Wiley & Sons, Ltd

*Other Wiley Editorial Offices*

John Wiley & Sons Inc., 111 River Street, Hoboken, NJ 07030, USA

Jossey-Bass, 989 Market Street, San Francisco, CA 94103-1741, USA

Wiley-VCH Verlag GmbH, Boschstr. 12, D-69469 Weinheim, Germany

John Wiley & Sons Australia Ltd, 42 McDougall Street, Milton, Queensland 4064, Australia

John Wiley & Sons (Asia) Pte Ltd, 2 Clementi Loop #02-01, Jin Xing Distripark, Singapore 129809

John Wiley & Sons Canada Ltd, 6045 Freemont Blvd, Mississauga, ONT, L5R 4J3, Canada

Wiley also publishes its books in a variety of electronic formats. Some content that appears in print may not be
available in electronic books.

*British Library Cataloguing in Publication Data*

A catalogue record for this book is available from the British Library

ISBN 978-0-470-03312-8 (HB)

Typeset in 10/12pt Times by Integra Software Services Pvt. Ltd, Pondicherry, India
Printed and bound in Great Britain by Antony Rowe Ltd, Chippenham, England.
This book is printed on acid-free paper responsibly manufactured from sustainable forestry in which
at least two trees are planted for each one used for paper production.

*To Katherine and Hannah, who will be choosing their own future over the next 20 years.*

# Contents

# Preface

My previous book forecasting the future, *The Future of Wireless Communications*, was published in 2001, but mostly written during 2000. This was the time just before the 'dot.com bubble' burst. At that point Internet and wireless companies were flying high, and to predict anything other than a bright future would have seemed highly pessimistic. Of course, that all changed during the latter part of 2000 and 2001 as is well known, and explored later. Now, predicting a bright future for wireless communications is more likely to be seen as unduly optimistic. As will be explored throughout this book, these changes in the market have broadly not changed the technological roadmap that we predicted earlier, but they have had an impact on the timing of the critical investment. Overall the events have been so tumultuous for the industry that another version of the book, some six years after the first edition, seems appropriate.

Along with the wireless communications environment, my role has also changed somewhat. In September of 2001, a few months after the publication of *The Future of Wireless Communications*, I left Motorola and returned to the UK and to the consultancy environment. After a spell with PA Consulting Group, the UK's sixth largest consulting company, I moved to Ofcom, the UK regulator of all things to do with communications, as the Head of Research and Development. This has given me the ability to commission research across a wide range of communications technologies and to be involved in developing policies, such as the UK's 'Spectrum Framework Review' that will shape the industry and the future in their own right.

Other appointments have helped me in this respect. I play a range of roles within the Institute of Engineering and Technology (IET) where I am a member of the Board of Trustees and chair of the *Communications Engineer* Magazine Advisory Panel. My appointment as a Fellow of the Royal Academy of Engineering provides me with access to most of the key engineers in the UK and my role as a Visiting Professor at Surrey University gives me access to world-class research. Finally, my appointment to a number of judging panels such as the *Wall Street Journal*'s Annual Innovations Awards has provided me with valuable information and contacts in understanding strategic changes in the industry.

In the preface of the earlier book, I noted that there seemed to be nobody producing a coherent and well thought-out vision as to how the whole world of wireless communications would develop, and perhaps converge, over the coming years; and that further, nobody seemed to be prepared to try to predict 20 years out. This still appears to be broadly the case – there have been a few books published with individual essays on the future but no coherent and complete text. Hence, I believe it is worthwhile to publish a new version of this book, to continue to provide some sort of a framework on which future strategic decisions can be made.

Those readers who recall the earlier version will remember that around half of the book comprised contributions from a range of eminent industry experts. I have used the same format in this edition, but have elected to ask a different set of experts for their views. I have made this decision in order to provide increased diversity of views and not because of any implied criticism of the previous experts – indeed, as will be demonstrated throughout this book they did an outstanding job in predicting the future.

Perhaps it is worth explaining why I am qualified to undertake such a task. I am an engineer, qualified to PhD level, but also with business qualifications. I have spent much of my career as a consultant, working across a wide range of different problems and issues with clients including regulators, governments, user groups, operators, manufacturers and SMEs. I have also spent time working for Motorola and for the UK regulator, Ofcom. Projects have varied from building hardware demonstrators of novel technologies through to developing business cases for operators planning national roll-outs. I have lived in the UK and the USA and worked for clients around the world including diverse environments such as South Africa, Bolivia, Denmark and Ireland. And above all else, I have previously made a range of predictions, almost all of which have proven to be accurate.

As always, it is important to remember that 'forecasting is very difficult, especially when it concerns the future'. The forecast presented in this book is highly unlikely to be correct in every respect. Regardless of this, the information presented here and the thought processes followed will be of value in helping others to build their own view of the future and to modify that view as circumstances change.

# Acknowledgements

Firstly, I am deeply indebted to those experts in the field of wireless communications who gave up their time to contribute substantial chapters and shorter pieces to this book. I would also like to express my thanks to the earlier contributors who made the previous version such a great book.

In writing this book I have drawn upon all my experience gained over my years in the industry. I have learnt something from almost everyone I have come into contact with and would thank all of those with whom I have had discussions. Special thanks are due to a number of key individuals. During my time at Multiple Access Communications, Professor Ray Steele, Professor Lajos Hanzo, Dr Ian Wassell and Dr John Williams amongst others have taught me much about the workings of mobile radio systems. At Smith System Engineering (now Detica), Richard Shenton, Dr Glyn Carter and Mike Shannon have provided valuable knowledge as have contacts with a number of others in the industry including Michel Mouly, Mike Watkins, Jim Norton and Phillipa Marks (Indepen). At Motorola I had tremendous guidance from a range of individuals including Sandra Cook, Raghu Rau, John Thode and the immense privilege of discussions with Bob Galvin, ex CEO. At PA Consulting, Dr Phil White, Dr John Buckley, Dr Mark Paxman and many others have contributed to my understanding of the cellular industry. In my work with Institutions I have been privileged to work with John Forrest CBE, Sir David Brown, Walter Tuttlebee, Peter Grant and many more. At Ofcom, Peter Ingram, Mike Goddard, those in my R&D team and others have provided invaluable guidance. Many presentations and papers from those involved in the mobile radio industry have contributed to my understanding.

Finally, as always, thanks to Alison, my wife, who supports all my endeavours to write books with good humour and understanding.

### *Disclaimer*

Note that the views and opinions presented in this book are those of the authors and not necessarily of the organisations that employ them. These views should in no way be assumed to imply any particular strategic direction or product development within the organisations thus represented.

# About the Author

William has worked in the wireless communications industry since his graduation in 1989. He initially joined Multiple Access Communications, a consultancy based in Southampton, UK, in 1989, becoming a director in 1992. There he worked on many research and development projects including detailed research into QAM for mobile radio – the subject of his PhD thesis – directing multiple propagation measurements, designing and writing a microcell propagation tool and producing hardware demonstrators for early DECT and QAM systems.

He moved to Smith System Engineering (now Detica), based in Guildford, in 1993. There he undertook a wide range of projects including taking the lead technology role on the 'Smith/NERA' studies into the economic value of spectrum and spectrum pricing. He played a key part in the standardisation of the European Railway's GSM technology, authoring many of the changes to the standard within ETSI and working across industry to build support for the proposals. He also worked on PMR systems, providing strategic and technical advice to major users such as police and fire services. While at Smith he completed his MBA, continued to publish widely and became well known as a speaker and chairman on the conference circuit.

In an effort to widen his understanding of the cellular industry, William then moved to Motorola. He spent a year in the UK working in the European infrastructure division, then he moved to Chicago in the United States where he was a Director of Corporate Strategy. He also worked closely with Motorola Labs on research directions. This exposure to strategic and technological issues enabled him to produce Motorola's medium- to long-term strategic plan. At this point he was elected a Fellow of the Institution of Engineering and Technology (IET; formerly the IEE) and had his biography included in *Who's Who in America*.

William then returned to consulting, moving to PA Consulting Group in Cambridge as a Managing Consulting in 2001. At PA he consulted widely across the wireless industry on a range of strategic, technical and regulatory issues. He was asked to become a judge for the *Wall Street Journal*'s Annual Innovation Award and has been invited to sit on this judging panel annually since. Also in 2003 he was invited to become a Visiting Professor at the Centre for Communications Systems Research at Surrey University. In 2004 he was elected as a Vice President of the IET where he sits on the Board of Trustees, directing the IET's strategy during a time of significant change and merger activity.

William joined Ofcom as Head of Research and Development just prior to it officially becoming the national telecoms regulator. Key outputs from the team include the Annual Technology Review and a range of research reports. William also leads across a wide range of spectrum strategy. He wrote the Spectrum Framework Review – Ofcom's long-term framework for radio spectrum management – and led on Ultra Wideband strategy and property rights for spectrum ('spectrum usage rights'). He is a member of Ofcom's Senior Management Group.

In 2005, at the age of 38, William was invited to become a Fellow of the Royal Academy of Engineering. This highly prestigious body comprises the top 1300 engineers in the UK. Admittance is by invitation only and limited to 50 engineers, across all disciplines, annually. Only a handful of individuals have ever been elected below the age of 40.

William continues to maintain a high profile in the industry. He has published over 70 papers in a mix of publications spanning learned journals to the *Wall Street Journal*. He chairs some six conferences a year, and speaks at another six or so. His books include: *Modern Quadrature Amplitude Modulation* published by John Wiley and the IEEE in 1994; *Introduction to Wireless Local Loop* published by Artech House in 1998; *Understanding Cellular Radio* published by Artech House in 1998; *The Complete Wireless Communications Professional* published by Artech House in 1999; *Single and Multi-carrier QAM* published by John Wiley in 2000; *Introduction to Wireless Local Loop – second edition: Broadband and Narrowband Systems* published by Artech House in 2000; and *The Future of Wireless Communications* published by Artech House in 2001. He has recently become a series editor for book series entitled 'Essentials in Wireless Communications' with Cambridge University Press. In 2006, his biography was included in Debrett's *People of Today*.

# 1

# Predicting the Future is a Necessary Part of Business

Almost all activities in the world of wireless communications require a forward-looking assessment. Operators who are deciding whether to buy spectrum at auction need to assess the likely services and revenue they can expect over the lifetime of their licence – often 20 years or more. Manufacturers need to decide which areas to focus their research activities on and which technologies and devices to develop into products. Academics and other researchers need to understand which areas will require the greatest advances and hence be most amenable to research. With the development of standards taking 5–10 years from inception to commercial product, those developing the standards need to predict what types of product will be needed and technologies available during the lifetime of their standard. There are many examples of poor forecasting – for example Iridium over-forecast the number of users who would be prepared to pay for an international satellite phone; and some examples of excellent forecasting, such as Vodafone's decision to enter the mobile communications marketplace when it was in its infancy. Getting these forecasts right is one of the most critical factors in building a successful business.

All of these players – manufacturers, operators, researchers, investors, regulators and more – make their own predictions of the future in developing their business plans or other business activities. In doing so, they sometimes draw upon available industry information such as analyst and consultant reports which forecast the uptake for specific technologies or services. But these often take a very narrow view of a particular segment or technology. As a result, they can forecast that a particular technology will be successful when compared to its direct competitors, but may not notice the advent of a disruptive technology which will change the entire competitive playing field. The track record of such forecasts, particularly over timescales of three years or more, is generally poor.

*Wireless Communications: The Future*   William Webb
© 2007 John Wiley & Sons, Ltd

It is the role of this book to provide a broad and long-term forecast, looking across all the different elements of wireless communications and into related areas such as wired communications where appropriate. This forecast can then become an overall roadmap under which those involved in wireless can build their more specific forecasts.

I prepared such a forecast in 2000: *The Future of Wireless Communications* (Artech House, 2001). Despite the tumultuous time that many communications companies have experienced between 2000 and 2005, the predictions have proved surprising accurate, as discussed in more detail in Chapter 2. Equally, much has changed since then. New technologies or standards such as WiMax and IEEE 802.22 are being developed. New approaches to using radio spectrum such as ultra-wideband and cognitive radio have been proposed. Services predicted in the previous edition, such as combined home and cellular systems, have actually started to be deployed.

The approach adopted in this book is very similar to the successful approach I adopted in 2000. The book starts with a careful analysis of the current situation, looking at underlying technological drivers, user demands, existing and emerging standards and technologies, and business drivers among the key players. It then turns to gaining views from a range of eminent experts in the field, ranging across different types of stakeholder and from different parts of the world. This range of views, coupled with the earlier analysis of constraints and drivers, is then brought together into a single vision of the world of wireless communications in 2011, 2016, 2021 and 2026. Such a single vision is much more valuable than a wide range of different scenarios which readers need to interpret and come to their own conclusion as to which, if any, they consider likely.

So, in summary, five years after my previous book, it is an appropriate time to have another look at the current environment, gather some fresh opinions and make another prediction 20 years into the future. This book provides a credible, broad prediction of the future which will aid all those involved in wireless communications in their most important task – making the right decisions themselves as to their best investments of time and effort.

# 2

# Previous Predictions have been Accurate

## 2.1 Introduction

The previous book made concrete forecasts for 2005, a period five years into the future from the time of writing. Just over five years later, it is possible to review how well we did and understand what this means for future predictions.

It is worth remembering that the period 2000–2005 has not been a happy one for many companies involved in wireless communications. Most major manufacturers and operators have seen their value decline considerably over this period and between them they have contributed to some of the largest write-offs in history. These corporate difficulties were not generally foreseen in 2000. The impact that they would have on wireless development might be expected to be profound.

## 2.2 There have been Huge Changes in the Telecoms Climate

*There was a huge erosion of revenue, jobs and confidence between 2000 and 2003*
Most readers will be only too well aware of the scale of the changes that took place in the telecoms industry between 2000 and 2003. Some will have been unfortunate enough to lose their jobs as a result, others to lose money on investments they have made. Those still employed became used to falling sales and budgets, to economy grade travel and to daily doses of bad news from the media. Take, for example, Lucent – one of the leaders in the telecoms industry in 2000 with over 100 000 employees and a share price hovering around $100. By mid 2003 it had fallen to 35 000 employees with further job cuts forecast and a share price of around $1.50. Just considering the major manufacturers and operators, it is clear that over 500 000 jobs were lost in the telecoms industry between 2000 and 2003.

*Wireless Communications: The Future*   William Webb
© 2007 John Wiley & Sons, Ltd

Of course, the pain was not restricted to the cellular industry. Across almost all sectors share prices fell by up to 50%. But technology-related companies were particularly badly hit, with the NASDAQ in 2003 standing at around 25% of its highs of 2000. Worst hit of all were the so-called dot.com start-up companies pioneering new ways of business over the Internet. By 2003, few of the many thousands of start-ups existed.

In the cellular industry there was a clear turning point in fortunes around the time of the 3G auctions in the UK and Germany. Between then and 2003, share prices of the major players – operators and manufacturers – fell steadily to around 25% of their peak values in 2000. Matters became sufficiently bad that the *Financial Times*, in April 2002, speculated that perhaps the cellular operators no longer had growth prospects and should be treated instead as a typical utility company, albeit one that was highly indebted.

The causes of the malaise in the cellular industry are generally well known. Operators could no longer rely on subscriber growth as penetration levels reached saturation in most developed countries. Add to this a slight decline in average revenues per user (ARPUs, normally measured on a monthly or annual basis) as tariffs fell and some operators were reaching a point where they were experiencing a slight decline in revenue. This situation, although undesirable, would not have been too bad if it were not for the general level of indebtedness of the operators at the time, resulting from a spate of acquisitions and the license fees paid for the 3G licenses – in particular in the UK and Germany.

The fees paid for the 3G licenses were predicted on significant growth in ARPUs as consumers increasingly used services such as web-browsing, video telephony, location-based services and a host of other offerings for which the operators predicated they would be prepared to pay much more than their current monthly bill. However, in 2003 indications were not promising. The first mobile data service – WAP – as has been widely discussed, was mostly unsuccessful, with many concluding that it was just too awkward and too expensive to browse the Internet on a mobile phone. The advent of a packet data service – GPRS – was predicted to improve data usage as it provided 'always-on' connectivity at higher speed and lower cost; but even though GPRS networks were available since 2001, in 2003 less than 1.5% of cellular phone users had a GPRS subscription. Worse, uptake of GPRS services was widely believed to be a strong indicator of the eventual demand for services on 3G. Elsewhere in Europe, operators that had launched offerings such as location-based services were reporting disappointing results.

Only in Japan had there been growth in ARPUs, initially through DoCoMo's i-mode service[1] and then through J-Phone's service that enabled photos to be sent as part of a text message. Whether this could be replicated elsewhere was unclear. On the one hand there is much about Japan's culture that is different, including long commute times where mobile data applications can relieve boredom, and a low penetration of 'conventional' fixed Internet making the mobile device the main browsing tool for many. On the other hand, applications such as sending pictures would appear to have interest regardless of culture.

The one clearly successful service in Europe was the use of short message service (SMS) where revenues by 2003 were in the region of 10–15% of voice revenues. SMS had proved profitable to operators not least because its data requirements were so low that it produced

---

[1] Broadly, this is mobile access to the Internet, however to sites where content has been carefully tailored for delivery to mobile devices.

no significant load onto the cellular network.[2] In 2003, many were pinning their hopes of a growth in ARPUs, particularly in the short term, on an enhancement to SMS known as the multi-media messaging service (MMS) which enabled attachments to be added to text messages. These might be pictures, ringtones or short music clips.

By 2003, the evidence suggested that adding new services to existing and enhanced cellular networks would not have the desired effects on ARPUs. This left the industry in crisis, with operators unwilling to invest because they were unsure that they would get any return on this investment, and preferring to repay debts instead. This lack of spending was devastating the supply chain – manufacturers, suppliers, consultants, etc. – which in turn tended to reduce the flow of innovative technologies and applications just at the time they were most needed. All told, in 2003, the future for the wireless industry was not looking bright.

### These problems were caused by overly optimistic growth in the late 1990s

Much has been written about what caused these enormous changes in the telecommunications industry. It is worth spending some time looking at them because understanding the root causes makes it easier to predict the likelihood of future repetitions.

First, it is worth understanding where the problems lie. If we look specifically at the cellular industry, the overall revenue flowing into the industry, in the form of payments from cellular subscribers, had not fallen. Indeed, for most operators it continued to grow throughout the period 2000–03, albeit at a lower rate of growth than in the heady 1990s. Such a slowdown was not unexpected as it was obvious that penetration levels would eventually reach saturation.

The problems appeared to reside on the supply side. Orders for equipment fell hugely compared to the late 1990s. So although the revenue entering the cellular industry had increased, it had not flowed out into the supply chain in the same manner as during the period prior to 2000. Instead, in many cases it was used for debt repayment. For example, in late 2002, France Telecom – owners of Orange – cancelled many capital projects to focus on debt repayment.

In tracing the cause of the debt, many point their fingers directly at the 3G auctions that took place around the world, but particularly in Europe, during 2000 and 2001. Certainly these contributed to the operators' debts – by some $100bn. However, they are far from the whole story – the debt of France Telecom alone stood at nearly $70bn in 2002. In practice, much of the debt actually arose from the takeovers which raged across the industry between 1998 and 2001 as the larger operators tried to establish a global presence. Others ran up debts as they invested heavily in operators with a minority share of the market in an often unsuccessful effort to increase their share.

But none of these are really the cause of the debts, they just explain where the money went. More important are the reasons why operators decided to spend the money in this fashion. These vary from operator to operator but include:

- a desire by some operators to gain a global presence under the belief that only the global players would survive
- optimism over the future of mobile data, leading operators to predict increasing ARPUs which then made the business cases for investment look attractive

---

[2] Indeed, on the basis of cost per bit, SMS is very attractive to the operators, being charged at around £1,000 per Mbyte!

- rapid growth in their share prices persuading them that they could increase debt without impacting debt to equity ratios.

The simplistic summary is that the industry, and in particular the operators, got too optimistic. Years of buoyant growth, high stock prices and the Internet bubble persuaded them to behave as if this growth would continue indefinitely. It is now clear that the growth slowed, rapidly exposing positions that were untenable in this kind of market.

### Confidence slowly returned to the industry between 2003 and 2005
Slowly, from around 2003, things improved. The levels of debt for the major operators declined substantially as operators repaid large amounts, or used other financial instruments to restructure and dilute the debt. No major operators collapsed, or even appeared to be in serious trouble, and the manufacturers started to return to profitability, albeit on a much reduced order base. Around 2003, 3G started to come of age, with major roll-outs from new entrants, such as 3 in the UK and Italy, and early deployments from existing 2G operators. This increased the overall spending in the industry, helping to restore confidence.

Operators accepted that ARPUs and subscriber levels were relatively static, hence overall revenue was static, and reorganised themselves accordingly, outsourcing, downsizing and generally optimising their internal structures for this new, more static role. There were some bright spots. For the new 3G operators, ARPUs appeared substantially higher than the industry average, raising hopes that the new data services offered by 3G would, after all, encourage users to spend more. Others were more cautious, noting that the offers put together by these operators were most attractive to the high ARPU spenders, so it was possible that migration of high ARPU users had occurred rather than users increasing spending levels.

New technologies started to become mentioned by 2005. For 3G, HSDPA was seen to be the next major advance. Outside of 3G, WiMax was widely expected to bring attractive new services to end users. Mobile TV also was trialled as a potential new service.

None of this changed the total revenue entering the industry which remained broadly static from 2000 to 2005. But the change in confidence lifted share prices of both operators and manufacturers and made all more inclined to innovate and try new ideas.

## 2.3 What we Predicted for the Period 2000–2005

In our previous book we predicted that during the period 2000 to 2005:

- Office buildings would increasingly have W-LANs composed of technology based on either BlueTooth, cellular pico-nodes or W-LAN standards.
- W-LAN coverage would start to appear in airports and some other public hotspots.
- Early versions of intelligent call filtering systems would be deployed in some networks but these would typically be restricted to devices operating on one network and would not be able to work across multiple networks.
- Early third-generation cellular networks would start to roll out.
- Cells and networks would become self-planning and optimising, reducing the management burden on the operator but increasing the software complexity delivered by the manufacturer.

We suggested that users would still have different phones for their office, home and mobile usage. Some offices might have integrated mobile phone systems but these would be relatively rare.

We thought that the core network functionality would be hardly any different from that utilised in 2000. Most core networks would still be based on conventional centralised switching platforms using signalling protocols such as SS7. Some limited Internet integration and third-party service provision might be provided and new third-generation operators might be utilising IP-based core networks, but the huge installed legacy base of Class 5 switches and MSCs would still carry the majority of the traffic. This would make it very difficult to connect different access pipes (such as cellular and fixed wireless) to the same core network and would severely limit the capability to have a single number and phone.

In the home, we predicted little change from the year 2000, with the exception that many more homes would have high-speed connections through fixed wireless, cable or ADSL, into the Internet. We thought that some homes would have implemented in-home networks so that multiple computers could be connected to the Internet, but these would have limited capability to communicate with many other devices.

We expected gradual penetration of BlueTooth during this period, resulting in the formation of some ad-hoc networks particularly between mobile phones and computing devices. This would allow automatic retrieval of emails using GPRS or third-generation equivalent services.

For the end users, we predicted that this period would see a gradual increase in data capabilities in mobile devices. Enhancements in simplicity would occur as BlueTooth and packet data networks enabled them to seamlessly retrieve data from pre-programmed points in the office or home network.

## 2.4 How Well did we do?

In short, very well. The predictions we made appear almost exactly right. In some cases we were unsure about which technology would transpire, for example whether W-LANs or cellular picocells would predominate, but the overall direction was correct. Interestingly, in overall terms we predicted little change of substance between 2000 and 2005, and that is exactly what transpired. This was not based on an expectation of hard times ahead for the wireless industry, but more on an understanding as to how long it would take for technologies to be developed and reach mass market penetration levels.

Were we lucky? Perhaps it might be argued that we under-predicted the amount of change that would have normally happened, but because of the downturn in the communications market, less happened than normal. Otherwise, how could we explain the fact that our predictions were correct even though we did not foresee the dramatic downturn in the market? It is a difficult judgement to make without observing the counter-factual situation where there was no downturn in the market. Without a downturn perhaps operators would have invested more quickly in new networks such as 3G and W-LAN. But regardless of investment amounts, development times tend to be similar and the rates at which consumers replace devices such as handsets and laptops relatively unchanged. So our verdict would be that while the downturn had traumatic effects for employees and shareholders in many companies, it actually made relatively little difference to the overall deployment of technology. Before the downturn, many companies had unrealistic plans, after the downturn they became more realistic.

## 2.5 Our Predictions for 2005–2010

So, given that we did so well for 2000–05, what of our predictions for 2005–10? In 2000 we predicted that during the period 2005 to 2010:

- Homes will start to deploy BlueTooth nodes.
- Communicator devices will be able to work on these in-home networks.
- Intelligent filtering and redirection functions will become widespread such that users will be able to have messages forwarded to them in the manner they wish, whatever network they happen to be connected to.
- Email and telephone systems will be linked together using mobile IP standards.
- Video communications will start to become widespread making up around 2% of calls.
- Broadband connections to homes, perhaps using wireless, will become common.
- Data rate requirements will reach around 10 Mbits/s.
- Speech recognition will become a standard means of communicating with machines.
- Machine-to-machine communication protocols will be widespread and robust towards the end of this period.
- Core IP networks for cellular systems will become commonplace.
- Public W-LANs in hotspots will become commonplace.

A number of key changes are predicted. Within the office, mobile phones will now be able to communicate with most networks. The office networks will no longer be connected back to the PSTN, but to a core IP network enabling other networks which also have a core IP base to communicate in a more seamless manner. However, not every cellular network will have moved over to an IP base and so not all will integrate seamlessly with office networks. This will mean that there will be some segmentation – business users will prefer the more advanced networks and will pay for the service, whereas some residential users will pay less for a lower functionality network.

Radio devices have also entered the home, adding to the in-home networks that some users had already implemented. These will start to work with some mobile devices, for the users that will pay additional fees for this capability. These users will now have removed their fixed phone. However, there may still be some difficulties in that the high-speed connection into the Internet may not allow all the functions required for unified messaging and other services.

Third-party intelligent call filtering and redirection systems will have started to appear. Coupled to these will be widespread third-party provision of services introduced through the Internet. Not all networks will work with these filtering systems and some operators will still be deploying their own systems. Filtering systems will be developing rapidly by this point, able to provide very innovative and personalised services to the end user.

Ad-hoc networks will be well developed, with PDAs, phones, computers and other similar devices routinely communicating with each other. Machine-to-machine communications will start to become more normal, but there will still be a lot of legacy machines and a lack of definition of languages preventing, for example, the fridge talking to the PDA in any meaningful manner.

For the end users, their requirements will be starting to be met. They will be able to have a single phone and a single number and complete unification of messaging if they select an appropriate set of networks and probably if they pay an additional fee for the provision of this service. They will be able to use video communications, although this will still be somewhat of a novelty. Their machines will start to communicate with them, although they will not be communicating with each other with the full functionality they might desire.

## 2.6 How Good do these Predictions Look Now?

We will examine these predictions in much more detail in the later chapters of this book. Suffice to say here that all these predictions still look reasonable. A few minor details now look wrong – home networks appear more likely to be based on W-LAN systems rather than Bluetooth and broadband connections are already quite widespread – but these appear to be the only changes needed.

What about WiMax, 802.20, UWB and many of the other developing technologies that have not been explicitly mentioned here, and were generally not foreseen in 2000? We think we have taken these into account, even though we did not know in exactly what form they would emerge. We have predicted that wireless might become one of the routes to the home – a key area for WiMax. We have suggested that Mobile IP might proliferate – the basis on which 802.20 standards are built. We suggested enhanced home networks – an area where UWB might play a role. Equally, in our earlier book we sound a note of caution. Both theory and experience suggest that there is unlikely to be a large gain in spectrum efficiency over 3G systems. The chance of a new technology leading to a revolution in wireless communications, rather than an evolution, seemed small.

## 2.7 Implications for Forecasting the Future

The key implication is that in the area of communications systems the future can be predicted accurately – at least five years out. This is contrary to popular thinking which assumes that wireless is such a fast moving world that predicting the future is near-impossible. Indeed, given that the 2010 forecast still appears broadly reasonable we would argue that predicting accurately ten years into the future and beyond is entirely plausible. The key is to base predictions on a sound understanding of technology, economics and perhaps most importantly the length of time it takes to develop standards and for devices to become accepted into the market. Most inaccurate predictions are overly optimistic as to the length of time it takes to introduce new technology. As our predictions have shown, it can take five to ten years for this to happen.

# 3

# How to put Together a Forecast

There are multiple interrelating measures that need to be considered in forming a forecast. Broadly these are:

- *Technology*. Will technical advances allow us to do more, to do things differently, or to do things less expensively?
- *User demand*. Will there be a demand coupled with a willingness to pay for the new things that are being offered?
- *Economics*. Can the technology be offered at a price that the users find acceptable?

There may be other factors that are particularly relevant in certain situations. Regulatory strategy can make a difference, perhaps by enabling a new service to be deployed which was previously prohibited. The ability to raise finance may be a key issue for some companies. Patent protection or other IPR issues may influence outcomes. And, of course, poor execution, at any stage, will reduce the chances of success. But these secondary factors will be of less concern to us unless it is apparent that one will have particular prominence in a certain area.

When forecasting, it is critical to consider each of these three major factors and their interplay with each other. For example, when considering WiMax we might firstly ask whether the technology is likely to deliver the data rates and range it claims. Having decided upon the performance figures that are most likely we need to ask whether there will be any user demand for a service with these parameters. If so, we can estimate the cost of deploying a network, build a business case, and then determine the likely tariff structure. We can then re-ask whether there is likely to be user demand given this cost. We may need to iterate – if we conclude that user demand might be lower than expected we may need to consider the impact of this on network design and economies of scale and re-forecast the tariff structure. This may lead into a spiral where it becomes clear there is not a point at which the network costs and the revenues balance sufficiently to make the service offering attractive. There may be some situations where the users themselves do not directly pay for the service – for

*Wireless Communications: The Future*   William Webb
© 2007 John Wiley & Sons, Ltd

example when using Google which is funded by advertisements. In this case, rather than the users' willingness to pay, the overall business model of the service provider needs to be understood.

Failures in forecasting often occur because:

- The capabilities of the technology are initially over-estimated.
- The user demand is assumed; or if users are surveyed, they are asked only whether they would like a particular service, and not what their willingness to pay might be.
- The economics of deployment are then not calculated.

Considering a wide range of previous forecasts it was clear that most were overly optimistic about the degree of change over time, particularly in the introduction of new services. The key historic exceptions were the forecasts made in the early 1980s for cellular penetration in the 1990s.[1] These were all far too low, assuming that cellphones would predominantly remain a tool for business.

The approach adopted in this book to assessing each of the three factors has been:

- Technology is assessed based on a mix of calculations of fundamental principles, observations of key trends and comparisons with other technologies leading to an understanding of the likely capabilities of new technologies.
- User demand is estimated from current usage, observed trends and an assessment of what has and has not worked in the past.

We have also made some use of the Delphi forecasting technique. This is where a group of experts are gathered together, asked for their views, and then each is given the chance to modify his or her view based on what was heard from others. Iterated sufficiently this process can reach consensus although we have deliberately not done that here so as to preserve differences between each of the contributions. Such differences are valuable in establishing the degree of uncertainty in the future.

It is often said that forecasters over-estimate the speed of technological introduction but under-estimate the impact that it will have. There is likely a strong element of truth in this. For this reason we have thought very carefully about the speed of introduction of new services and often deliberately taken what appears at the moment to be a conservative view.

The following four chapters look at the current position in terms of available services, the likely changes in user demand in the future, technological progress and the potential impact of major changes in the world. These form a precursor to the views of the experts. In the final sections all these elements are brought together to deliver a future prediction.

---

[1] This was explored in some detail in our 2001 book.

# 4

# The Current Position

## 4.1 The Value of a Good Understanding of the Starting Position

In order to predict where we might be 20 years from now, it is essential to understand in detail where we are today, and in particular, what programmes are in place that will set the direction for the next 5–10 years. Some processes such as standardisation and spectrum allocation take from 5 to 15 years. So, in some cases, fairly accurate predictions at least of what technology will be available can be made for up to 15 years from now by understanding where we are today and the spectrum allocation programmes that have been set into place. This is sufficient to understand many of the possibilities but not necessarily the outcomes since it does not take into account factors such as user acceptance – discussed in the next chapter.

There is a wide range of wireless devices used for communications purposes. Today, these can broadly be characterised as mobile, fixed and short-range. Broadcasting is a somewhat related topic that we will also touch upon. The main divisions within these categories are listed below.

- *Mobile:*

  - Cellular

    - 2G
    - 3G
    - So-called '4G'

  - Private mobile radio

    - Analogue
    - Digital

*Wireless Communications: The Future*  William Webb
© 2007 John Wiley & Sons, Ltd

- Mobile mesh
- Emerging technologies including cognitive radio and software defined radio (SDR)

- *Fixed:*

  - Point-to-point
  - Point-to-multipoint
  - Fixed mesh

- *Short-range:*

  - W-LANs

    - A range of members of the 802.11 family
    - Zigbee

  - W-PANs

    - BlueTooth
    - High-speed variants such as WiMedia/UWB

  - RFIDs

    - Passive devices
    - Active devices

- *Broadcasting:*

  - Conventional analogue and digital broadcasting
  - Mobile broadcasting to handhelds.

In this section, each of these categories is examined along with a short discussion of networks and protocols.


## 4.2 Mobile Communications

### 4.2.1 Cellular

**Introduction to Cellular [1–3]**

By 2006 most cellular operators had started the process of deploying 3G, and in some cases gradually migrating customers from their 2G networks to 3G. This promises to be a long process with 2G and 3G networks running in parallel for many years to come.

The key developments on the revenue side are:

- Increasing penetration, to the level where there were more mobile phones than fixed phones in many countries. This resulted in slow growth in subscriber numbers, and in some cases falling numbers as the less profitable subscribers were persuaded to leave the network.

- Stagnating revenue per user as the cost per minute of a call fell with increasing competition but few other new revenue streams proved popular.
- The growth of the 'mobile virtual network operator' (MVNO) where a company markets a cellular service but does not own a network. Instead they buy wholesale network capacity from 'network operating companies' and resell it to subscribers. This is being tried in many different forms in different countries with some models proving successful.
- Significant promotional activity around packages comprising picture messaging, ring tone download, gaming and other mostly recreational services.
- A continual search for more revenue streams including picture message (which to date has been unsuccessful), web access (which is slowly growing), gaming (again mostly unsuccessful) and, by 2006, mobile TV. Despite all these efforts, ARPUs are constant or falling.

On the technology and network side, the following topics are being discussed:

- the extent of deployment of 3G and the use of enhancements such as high speed downlink packet access (HSPDA)
- the role of other technologies such as WiMax
- the ways of achieving in-building coverage
- the use of convergence technologies such as BT Fusion.


**2G Cellular**

Second-generation cellular systems are well-established, with over two billion users throughout the world. Technologies that qualify as 2G systems include GSM, CdmaOne and TDMA. Although capabilities vary between technologies and continue to evolve, the key attributes of 2G systems are typically:

- digital transmission
- voice capabilities
- circuit switched data capabilities up to around 60 kbits/s
- increasingly, packet switched data capabilities also up to around 60 kbits/s.

*2G is continuing to evolve.* 2G systems, particularly GSM, have come a long way from their initial launch in 1992. At that point they provided voice and limited data capabilities. By 2006 a host of other features had been added including packet data using the general packet radio service (GPRS), multi-media messaging service (MMS), high-speed circuit switched data (HSCSD), workgroup capabilities like group calls, protocols for browsing such as the wireless access protocol (WAP), higher data rates through Enhanced Data rates for GSM Evolution (EDGE) providing up to around 300 kbits/s in some situations, and many other features. This has meant that although it is now over 20 years since design work started on GSM, it has remained a generally good solution for the communications needs of its users. The concept of lengthening time between generations of cellular technology being bridged through evolution of standards is now becoming increasingly accepted. This has the implications that moving to a new generation is not as urgent a need as it has been in the past, and indeed may be more difficult as it is harder for the new generation to offer a major improvement in functionality.

*2G will be here for some time.* When 2G was deployed it was envisaged that 2G systems might have a lifetime of around 20 years. Many of the licences given to operators reflected this, with end dates ranging from 2008 to 2020, but with many clustered around 2010–2015. When setting this end date it was thought that the spectrum might be 're-farmed' for 3G usage – and indeed this thinking still persists widely among many regulatory bodies. However, it now appears that complete re-farming is unlikely for some considerable time. First, 3G systems are taking longer to introduce than previously expected (see below), so that even if 3G were to entirely replace 2G, this might take another 10 years. Second, most operators now expect to deploy 3G in cities and along major communication corridors, but not in the more rural areas where coverage will continue to be provided by 2G for some time. Third, operators see 2G as being the 'roaming standard' for quite some time since, regardless of which technology their local operator has deployed, most handsets will retain 2G capabilities. Serving roaming customers is relatively lucrative and not something the operators would wish to forego. It is likely that, in 2015, 2G will still be widely deployed.

*Indoor solutions struggle to get traction.* While 2G systems have had great success in providing wireless communications outside of buildings and in providing coverage inside the building based on signals penetrating from outside, their success in dedicated in-building deployments has been much more limited. Many attempts have been made by manufacturers to deploy 2G in-building solutions based on a range of different technologies, but these have not had any real success. There have been a range of reasons for this, including:

- Reasonable coverage inside many buildings is already achieved simply from signal penetrating the building from outside, reducing the need for a dedicated in-building solution.
- Building owners cannot deploy their own 2G solution as they do not own spectrum, but they are typically unwilling to allow a cellular operator to deploy a solution and hence become the monopoly supplier.
- GSM is not well-suited to in-building deployment, with limited capacity, frequency planning problems and handover issues.
- The cost of cellular is still seen as higher and the quality as lower than fixed line communications.

In the UK in 2006 there was another initiative under way to address this space. The regulator, Ofcom, auctioned the so-called 'DECT guard bands'. These are a small piece of spectrum between the top of the 1800 MHz GSM bands and the bottom of the 1900 MHz DECT band. It had been thought this guard band was needed to avoid interference, but by 2005 it was clear that this was no longer the case. A key advantage of this band is that existing 2G handsets are able to use the spectrum without modification, allowing immediate access to a potentially large customer base. Possible applications include campus networks, office networks or home networks. However, there remain substantial difficulties such as automating the process of phones roaming on to and off these local networks, making it likely that any progress in this space will be slow, and likely lagging behind convergent solutions such as the BT Fusion product discussed below.

Given that attempts have been made for many years to deploy 2G in-building, and that no major technological breakthroughs are foreseen, it seems reasonable to assume that 2G will not be deployed in the majority of buildings. The balance of evidence and opinions suggest that instead it is more likely that a short range wireless solution will be deployed – something we discuss more in later chapters.

## 3G Cellular

In 2006, 3G systems were starting to be widely deployed.[1] The picture differed from country to country and even between operators, with new operators predictably having the largest deployments. There is a range of technologies for 3G, all based around CDMA technology and including W-CDMA (with both FDD and TDD variants), TD-SCDMA (the Chinese standard) and Cdma2000 (the US standard). The key characteristics of 3G systems include the ability to carry video calls and video streaming material and realistic data rates extending up to 384 kbits/s in both packet and circuit switched modes.[2]

For some time there was heated discussion about the likelihood of 3G being successful. With slow deployment and limited subscriber growth between 2000 and 2004 there were many who were quick to predict disaster. Others pointed to the growth of WiFi and the imminent arrival of WiMax and predicted that this would significantly disrupt 3G deployment. However, between 2004 and 2006 these doubts slowly faded. Operators continued to deploy 3G networks and subscriber numbers started to rise more quickly. Handsets became more attractive and the lower cost per call of 3G compared to 2G became increasingly clear. In 2006, few would now doubt that 3G networks will be widely deployed, that they will eventually take over from 2G in most areas and that subscribers will increasingly migrate from 2G. However, growth may occur more because the cellular operators 'push' the new technology than the subscribers demand it. Many subscribers may find themselves migrated over to 3G when they require a new phone despite the fact that they do not want to use any of the new services offered. It is possible that operators will not make a good return on their total investment comprising auction fees, network deployment costs and handset subsidies. However, we will eventually all be using 3G. Whether 3G can be judged a success under such conditions depends on your point of view!

*How things have changed.* The original premise of 3G was that end users would like to be able to do much more than just voice and low-speed data with their mobile phones. In addition, they might like to have video calls, watch video clips such as key sporting highlights, transfer volumes of data requiring more than 100 kbits/s, browse websites when outdoors and much more. It was suggested that 2G systems were unable to do this because they had an insufficiently high data rate for some applications and insufficient capacity to add video calls and other new applications on to the network. Early launches of 3G networks did indeed stress the ability to have video calls, see video clips and so on. However, in 2006, the situation was that by and large the benefits of 3G were being realised as increased voice call capacity, allowing more calls for less cost. Video calling capabilities were not being

---

[1] Because W-CDMA is a major change from GSM, in the European standard arena it is simple to label W-CDMA as a 3G technology. However, Cdma2000 has evolved from CdmaOne via a number of steps and so it is less clear at which point 2G stops and 3G starts. There is no clear dividing line that can be used, as a matter of somewhat arbitrary convention we will assume that systems labelled as Cdma-1X data only (DO) and data and voice (DV) systems are 3G while cdma-1X is 2.5G.

[2] 3G systems were designed against a criteria of delivering 2 Mbits/s data rates. Most 3G systems are technically able to do this but for many practical reasons will not go beyond around 400 kbits/s without enhancements.

widely promoted, instead downloads of music videos were offered. So far, there appears to have been a major disconnect between the initial view of what users would want from 3G and the experience of what users actually paid for. It may, however, be too early to judge this. Major changes in behaviour often take some time. For example, the Internet bubble of the late 1990s was based upon the assumption of, for example, mass purchasing via the Internet. This failed to transpire by 2000 resulting in the bubble bursting. But by 2006 Internet purchasing had reached some 10% of total purchasing and it was clear that a major shift in purchasing patterns had occurred. Following this parallel, it may take ten years from the introduction of 3G before some of the services it offers become widely accepted.

*The outlook for 3G.* As discussed above, 3G networks will be deployed by virtually all of the major operators and subscribers will be transferred over to the new networks. 3G will probably have a lifetime similar to that of 2G – in the vicinity of 20 years. With enhancements to the standard this lifetime could be even longer.

3G will face competition from other technologies. W-LAN systems will continue to proliferate in hotspots, homes and offices. Indoor voice and data traffic will increasingly migrate to these W-LAN networks which will be cheaper, faster and probably provide better signal quality. WiMax may provide competition, although, as discussed below, this is not certain. The use of new services may also expand on 3G as consumers become more comfortable with them, or as penetration grows. Growing penetration is key for two-way services such as video calls which require both parties to have a suitably equipped phone.

## WiMax for Mobile Applications

Around 2004, WiMax (or IEEE 802.16) became increasingly discussed. Although initially designed as a fixed technology (see below), it was being adapted in the standards to become a mobile technology (also known as 802.16e). Major claims were made for WiMax. Some claimed it would have significantly greater range than 3G, higher data rates, would be built into all computing devices and as a result would dominate the market for 3G-like services.

In 2006, the view was that WiMax mobile networks might not be launched until 2007 or 2008. It would then take some time to deploy the networks and seed terminal devices, so it might not be until 2010 that it would be possible to see whether WiMax was having any impact on 3G. Nevertheless, some simple observations suggest that its impact will be limited.

- *Range.* The range of a technology is typically dominated by the frequency of operation, the power transmitted and the height of the transmitter and receiver. All but the frequency are likely to be similar between 3G and WiMax. Although the frequency for deployment of WiMax is not certain, the most likely band at 3.5 GHz is higher in frequency than the 3G bands at around 2.1 GHz. Range will, as a result, be lower – perhaps somewhere between 50% and 75% of the range of 3G.
- *Data rates.* The technology used for WiMax (OFDM) is not significantly more spectrally efficient than the technology used for 3G (W-CDMA). (This is discussed further in the appendix to this chapter.) However, OFDM coupled with a high channel bandwidth will allow greater peak per-user data rates. So, on average, for an equivalent spectrum allocation, users will see similar data rates. In specific situations, where there are few users, it is possible that WiMax will provide a higher data rate. However, in commercial systems, such situations are likely rare.

- *Cost.* The network costs of WiMax will likely be higher than for 3G because of the reduced range and hence the need to build more cells. The subscriber subsidy costs may be lower if WiMax is built into processor chips, although this may not apply if users wish to have WiMax handsets.
- *Timing.* WiMax will arrive some five years after 3G is well established. This disadvantage in time is likely to be significant since without a compelling advantage few will choose to move from 3G to WiMax. However, those yet to deploy a system may find the choice balanced between the two technologies.

Hence, our view is that WiMax will not have a major impact on the success of 3G. It is most likely to be deployed by those operators who have not yet rolled out 3G networks.

## '4G' Cellular

With the introduction of 1G in 1982, 2G in 1992 and 3G around 2004, the deployment of 4G some time around 2014–2018 might look like a fairly certain bet. Indeed, this was one of the predictions made in the earlier version of this book. In the last few years there has been much discussion around 4G, with conferences and books published. Some thought that it was appropriate to start discussion of 4G in 2002 given the likely ten years it would take to complete the standard and have equipment developed. An example of this was the Japanese authorities who announced a research programme aimed at producing a system capable of delivering 100 Mbits/s to end users. Others argued that 3G had failed, or was inappropriate, and 4G should be introduced rapidly in its place. As might be expected, some of the key proponents of a rapid introduction of 4G were manufacturers with product that they classified as being 4G. However, during the early part of 2003, the ITU, mindful perhaps of the slow and somewhat uncertain introduction of 3G, decided to put on hold any discussions about 4G systems, thus delaying the introduction of 4G to a date of 2015, if not later.

In Chapter 6, the technologies that might be used for 4G are discussed in more detail and the conclusion is reached that 4G will not be a new air interface, unlike 3G. Even without examining the technical issues, Figure 4.1 suggests why 4G is likely to be different from 3G, and indeed perhaps may not even emerge.

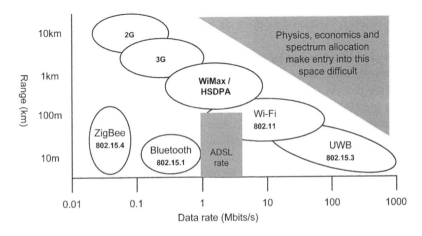

**Figure 4.1**   Range versus data rates for each of the generations.

Figure 4.1 illustrates that there is always a trade-off between range and data rate. Broadly, higher data rates require more spectrum. More spectrum can only be found higher in the frequency band but higher frequency signals have a lower propagation range. Each generation has accepted a shorter range in return for a higher data rate. But, as Figure 4.1 shows, the next 'step' in the process, where 4G might logically fit, is already taken by a mix of 3G enhancements, WiMax and WiFi. Indeed, interestingly, the Japanese plans for a 4G technology talk of an OFDM-based solution in the 3–6 GHz band providing up to 100 Mbits/s of data. This is almost identical to the specification for the latest W-LAN system, 802.11a, which is already available and exceeded by proposals for 802.11n.

Figure 4.1 also tells us much about the structure of future communications networks. Higher data rate systems are preferred but with their shorter range can be economically deployed only in high-density areas. So we might expect a network where 2G is used to cover rural areas, 3G to cover urban and suburban areas along with key transport corridors, and W-LANs providing very high data rates in selected 'hotspot' locations such as airports.

Finally, Figure 4.1 also illustrates another constraint – that of backhaul. Many WiFi cells are connected back into the network using ADSL. This has data rates in the region of 1–2 Mbits/s at present, much lower than the rate that the W-LAN air interface can provide. A further increase in air interface data rate in such a situation is pointless. Instead, research needs to be focused on better types of backhaul. This is discussed in more detail in Chapter 6.

Hence, an argument that will continue to be developed in subsequent chapters is there may not be another generation of cellular technology because there may not be sufficient economic justification for the development of a completely new standard. Instead, we might expect to see enhancements to all the different standards making up the complete communications network, with perhaps some new niche standards emerging in areas such as ultra-short range communications. This has not stopped some claiming that they already have a 4G technology, or like the Japanese, that they are working on one. Of course, since there is no widely agreed definition of what comprises a new generation, anyone is free to claim that their technology meets whatever criteria they regard as important, and with the definition of 3G already confused, perhaps we should expect 4G to be even more opaque!

To some degree, the relentless march of the next generation has provided a framework for the entire industry to orchestrate all the factors needed to progress cellular communications. With the end of this regular process there are a number of key implications.

- The industry has relied on regular injections of new spectrum. Without a new generation to pull this through, regulators will need alternative mechanisms. It is possible that spectrum trading, in the process of being introduced around the world in 2006, might provide this solution.
- Much research and development (R&D) over the last 20 years has been focused on more efficient mechanisms to transmit information. This is now increasingly hitting the laws of diminishing returns.
- Conversely, as the number of different technologies being used grows the complexity of making them all work together increases.
- The manufacturing industry has relied on new generations to stimulate spending throughout the industry. The end of the generation game will require them to adopt a different strategy.

As a final conclusion on the developments taking place in 2006, many of the proposed '4G' solutions were based on OFDM modulation. Claims were being made that OFDM was the new modulation of choice for 4G in the way that CDMA had become for 3G. OFDM is discussed in more detail in the Appendix to this chapter where it is concluded that it does not offer significant improvements in performance, but may be a pragmatic solution under certain circumstances. Therefore, we are not of the view that there needs to be a change to from CDMA to OFDM in order to realise significant gains in throughput. Equally, we are not suggesting that OFDM is inferior to other technologies, it just offers a different mix of trade-offs.

## Convergent Technologies

Within the wireless industry, and indeed as postulated in the earlier version of this book, there remains a view that a single handset that worked in the home, office and wide area would provide value to the end user. To some degree, cellular does provide coverage in these environments and the number of people having only a cellular phone and not a home phone has been steadily growing. However, there are problems with using a cellular phone in the home or office.

- Coverage throughout buildings is not always adequate.
- Quality is sometimes poor. Many prefer the quality of a landline to that of a cellular phone.
- Call costs can be higher for cellular phones.

There have been multiple attempts to address this. Initiatives to provide cellular 'home base stations' have been tried with GSM and are currently being developed for 3G. In the past, these have failed because of the cost and the complexity of radio spectrum management and handover management with potentially millions of base stations. Initiatives to use other technologies were also tried; for example BT offered a dual GSM/DECT phone in the mid 1990s. However, the phone was bulky and the offering was primarily aimed at the enterprise, which did not appear to appreciate the benefits. In 2005, BT made another attempt to target this market with 'Fusion'. This comprised a dual-mode GSM/BlueTooth handset and a BlueTooth node for the home. It has been marketed with a wider range of handsets which do not have a size penalty. It is still too early to say whether this will be successful. BT has also led the development of a standard for this kind of access, known as unlicensed mobile access (UMA).

Our view is that W-LAN is steadily becoming the de-facto standard for home networks. Users are increasingly deploying W-LAN to interconnect computers, allow laptops to roam around the house, and in the future to interconnect other devices such as security systems. Most people who deploy a W-LAN connect it to a broadband pipe to the home. A handset that can work on cellular networks and then hand over to W-LAN when in the home would seem to offer some advantages. Many manufacturers are now working on such handsets; and, as applications such as VoIP over broadband become widespread, we expect such handsets to become increasingly popular. Hence, we expect W-LAN in the home and office coupled with cellular in the wide area to become the convergent technology, gaining increasing penetration over the next ten years or so. This is discussed further in the last part of this book.

**Summary for Cellular**

The cellular industry has passed through a difficult period from 2000 to 2005. This now appears to be behind them, with almost all operators reporting significant profits. The operators are all also well down the road of deploying 3G and have a fairly clear strategic roadmap ahead of them. There appear to be few threats to cellular revenues, with the exception of in-building voice calls transferring to W-LAN over time. The likelihood of dramatically new or destabilising technologies appears low. Hence, we predict a long period of stability, with profitable operation and deployment of 3G.

## 4.2.2 Private Mobile Radio

**Introduction**

Private mobile radio (PMR) is one of the oldest forms of wireless communications. In its simplest form a company deploys its own radio system to assist in its operations. For example, at one extreme, a taxi company might erect a mast on the roof of its premises to broadcast to its taxis in the vicinity. At the other extreme, a national police force might deploy its own national network to provide communications to its staff throughout the country. There are a range of reasons for operating a PMR system.

- *Cost.* For operations such as taxi companies, the cost of deploying their own system is much less than the cost of making calls over cellular. However, this only holds true where local coverage is needed. When coverage needs to be more widespread, the cost of the PMR system grows while that of using cellular stays the same.
- *Features.* For organisations such as emergency services, features like group calls and fast call setup are considered essential. Generally, such features are not available over cellular, and so PMR solutions are needed.
- *Certainty of access.* Only by owning their own system and radio spectrum can an organisation be absolutely sure that calls will get through at critical times. This is especially important where safety of life issues are at stake (e.g. with the railways), but is also of economic and strategic importance to a business.
- *History.* Organisations that have operated their own system in the past, perhaps before cellular technology was widespread, often find it hard to change because of the logistics of swapping mobiles, control centres, IT systems, etc. Further, individuals in the communications department might perceive that a move away from PMR would eventually result in job losses.

In the last five years, there have been a number of forces decreasing the attractiveness of PMR, including:

- *Cost of spectrum.* Some countries, like the UK, have started charging for the use of the radio spectrum at a level designed to be close to its predicted value under auction. This has often had a significant impact on the cost of running a PMR system. However, its impact on the cellular operators is negligible because their spectrum is shared among a much larger number of users.

- *Enhancements to cellular.* As cellular systems have added packet data capabilities, enhanced coverage and become stable and dependable, some of the barriers to using cellular have disappeared.
- *Trends toward outsourcing.* The general trend to outsource non-core activities in order to save money has resulted in pressure to move from self-provision.

Figures for PMR usage are extremely hard to find on a world-wide basis – often because the information is not collected. Where they are available, they suggest that user numbers for PMR are falling by around 5% per year. On this basis, PMR will be a declining but still important part of wireless communications for many years to come. However, because its user group is relatively self-contained, and generally slow to change, it seems unlikely that it will be a driving force in the future direction for wireless.

The sections below provide a brief description of the two key classes of PMR – analogue and digital communications.

**Analogue Communications**

Organisations like taxi companies typically use analogue PMR systems. At their simplest, these systems consist of a single channel to which all the mobiles are permanently tuned and on which anyone can respond at any time. These solutions are often effective because they provide broadcast calls which fit well with the mode of working of organisations like taxi companies.

More complex variants of analogue systems add the capabilities to work across multiple channels (often termed trunking) to increase capacity, and in some cases the ability to overlay data on to the analogue transmission (at which point the distinction between an analogue and a digital system starts to get somewhat blurred). Most of these types of system conform to a UK standard known as MPT1327. As an example, in 2005, London Buses, the largest bus fleet in the world, was using this radio system to provide voice and data communications to their 7000 buses. The system enabled them to:

- broadcast voice messages to all, or subsets of all, buses
- receive voice calls from drivers, with emergency pre-emption if needed
- receive automatic vehicle location (AVL) data from the buses.

To achieve this, London Buses had needed to make a number of proprietary modifications to their system. This system has served them well for many years, but by 2005 it was starting to suffer from capacity problems and was not being well supported by major manufacturers.

**Digital Communications**

Digital PMR systems have been available since around 1995. Key solutions include the European standard TETRA, the US standard APCO25 and the proprietary TETRAPOL. Digital systems offer advantages.

- *Enhanced capacity.* By using voice coders to compress voice, more users can be squeezed into the same spectrum – typically about twice as many.

- *More features.* Additional features include security, packet data transmission, more advanced calling mechanisms and better quality of transmission.

However, digital systems have a number of disadvantages including:

- *Cost.* They are more expensive than their analogue replacements – sometimes by a factor of two or more.
- *Spectrum.* They are typically available in only a limited number of spectrum bands – often a subset of the bands for which analogue equipment can be purchased – making it difficult for some users to migrate from analogue to digital.
- *Scale.* Most digital systems are designed to be used with a number of base stations and a central controller. They often do not function well, or are very expensive, if used in a single site configuration.

For some larger user groups, these advantages are overwhelming. For example, for police forces, security is a critical issue and well worth the extra cost of a digital solution. For others, such as London Buses, the enhanced capacity is a bonus but otherwise digital brings limited benefits.

During the last decade, many countries commenced the process of rolling out national digital PMR networks for their emergency services. This seems likely to be the key application for digital PMR[3] and these systems seem likely to remain in deployment until at least 2025. However, also during this period national deployments of digital PMR systems for commercial users were attempted – for example in the UK under the name of Dolphin. These systems have failed to attract anything like the subscriber numbers they needed for profitability[4] and as a result generally went into administration.

PMR systems will continue to inhabit their own 'space' in the wireless communications market, with limited influence on what happens elsewhere, and with a slowly decaying number of users.

### 4.2.3 Mobile Mesh Systems

In 2005, mesh technology was much discussed. Predominantly, this was in relation to a WiFi network or 'cloud' being widely deployed across a city, with many of the WiFi nodes being interconnected using a mesh network. There was also some discussion of true mesh networks where mobile nodes communicated directly with other mobile nodes, mostly in the context of cognitive radio.

The concepts behind mesh technology and its likely success are discussed in Chapter 6 and not mentioned further here. If successful, a WiFi 'cloud' across a city could be a potential replacement for cellular, effectively allowing convergence to extend from the home and office into the urban area. However, it currently appears unlikely that such mesh systems would have the capacity to support large numbers of voice calls unless most cells were connected via high-speed backhaul links. At this point the costs of the system become similar

---

[3] Indeed, the name TETRAPOL, derived from 'TETRA for police' hints strongly at this expectation.
[4] Dolphin was aiming to attract around 1.5m customers. In 2003, when they went into receivership, their customer base was around 35 000.

to those of a microcellular 3G system, negating one of the key reasons for its deployment. Hence, we do not believe that mobile mesh networks will be a significant threat to systems such as 3G, although they may be successful in a range of niche applications.

## 4.2.4 Cognitive Radio

Another concept in much discussion in 2005 was cognitive radio. Although there were many different definitions of what a cognitive radio was, the basic concept was of a device which, on arriving in a new environment, would 'understand' the usage of the radio spectrum and adapt its behaviour accordingly. It was particularly thought useful for exploiting under-used spectrum. So, for example, a cognitive radio might detect that the emergency service frequencies were currently lightly used – perhaps because there were few emergencies taking place. It might then move to these frequencies and make a series of short transmissions, checking after each one that the frequencies were still free. If not, it might then move to other frequencies, perhaps those used by broadcasting for example, that were essentially unused in the area.

There are clear military benefits from such a system – it allows devices on a battlefield to work without central planning while taking into account existing local usage of radio. In the commercial world, cognitive radio holds the promise of expanding the amount of radio spectrum available for communications by making use of other spectrum. Measurements have suggested that much of the spectrum is lightly used so it is possible that a very significant expansion in the 'effective' amount of spectrum available for personal communications might be achieved. In understanding the value of cognitive radio we need to ask whether it will work, and how valuable this additional spectrum might be.

*Will it work?* Our view is that it will not work. One of the key problem for cognitive radios is the so-called hidden terminal problem. One possible way this might occur is illustrated in Figure 4.2.

A cognitive radio user might make a measurement and not spot any activity on a piece of spectrum. However, there might be a legitimate user of that spectrum behind the next building, transmitting to a tower on the hill. Because the building is between the users, the cognitive radio user does not receive the legitimate signal and so concludes the spectrum is unoccupied. But because both users are visible to the tower on the hill, when the cognitive radio user transmits its signal it is received as interference at the tower. Hidden terminal problems might also occur in many other situations, including the case where there is no shadowing but the cognitive radio is simply less sensitive than the legitimate receiver.

This problem is solved by the tower on the hill transmitting a signal, or 'beacon', indicating whether the spectrum is free. A terminal then requests usage of the spectrum, and if granted, the tower indicates that the spectrum is busy. Such an approach works well but it requires central management by the owner of the band. Hence, it becomes a choice of the owner of the spectrum as to whether they wish to allow this kind of access and if so under what conditions.

The hidden terminal problem is reduced somewhat for the so-called 'broadcasting white space'. TV broadcasters will use a particular channel in one city. They may not use the same channel in an adjacent city to avoid interference, only reusing it perhaps in a location hundreds of miles away. In between these two transmitters there may be a chance to use the frequency at a much lower power level than the broadcasting transmitter without causing

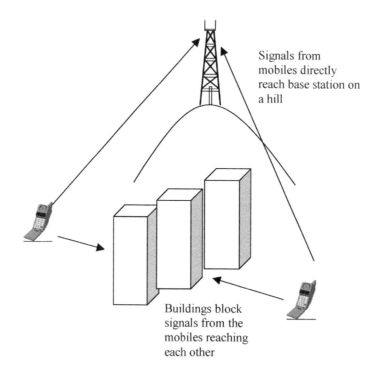

Signals from
mobiles directly
reach base station on
a hill

Buildings block
signals from the
mobiles reaching
each other

**Figure 4.2**  The hidden terminal problem.

interference. Because broadcasting use is relatively static – frequency allocation is rarely changed – terminals could determine from their location and a database whether the spectrum is usable, overcoming the hidden terminal issue. The extent to which there is broadcasting white space varies. In the USA, with its widely dispersed cities, it is prevalent. In the UK, with a more homogeneous population distribution and active use of the white space by programme making and outside broadcast equipment, there are fewer opportunities.

Other problems exist for cognitive radio, including the fact that a cognitive radio's emissions might spread into adjacent bands that were not scanned by the radio, causing interference into neighbouring users.

Even if all these problems could be overcome, there is still a question as to what the cognitive radio transmits to. A network of base stations which is able to scan all frequencies for possible transmissions would be expensive and hugely risky to construct. Locating other terminals to transmit directly to is also difficult and suffers the problems associated with mesh networks.

*Is the spectrum needed?* Our view is that there is little need for this additional spectrum. 3G operators in 2005 were still typically only using 50% of their spectrum allocation. Additional 3G spectrum was promised at 2.5–2.7 GHz and at UHF after analogue TV switch-off. As discussed later, cellular demand may eventually fall as more traffic flows to W-LANs and there is plenty of spectrum for these at 5–6 GHz. So, while shortage of spectrum has been an issue in the past, we believe this issue is decreasing.

Hence, we believe that cognitive radio will not be able to deliver on the promise of providing interference-free usage of under-utilised spectrum and will struggle to find an application where its more expensive handsets can be justified. It will not make a significant impact on our predictions for the future.

## 4.3  Fixed Wireless

### 4.3.1  Introduction

Fixed wireless [4] became very popular around 1997, with many postulating that it would be the next cellular boom. This failed to materialise with the result that, between 1999 and 2003, many fixed wireless manufacturers and operators left the industry. Even so, there continues to be a stream of new technologies and new operators entering the market.

The basic concept of fixed wireless is to provide a connection from the Public Switched Telephone Network (PSTN) or Public Data Network (PDN) to the home or business. Fixed wireless might be used for:

- *Simple voice communications to the home*, providing a service equivalent to the fixed line service that many developed countries already provide to most homes. This is often termed plain old telephony service (POTS) and includes voice, fax, data at up to 56 kbits/s and the provision of customised local access supplementary services (CLASS) – capabilities such as emergency calling, three-party calling and other supplementary services.
- *Enhanced voice and data communications*, providing all of the above features, more than one simultaneous voice line plus enhanced data capabilities. Typical offerings in this area at the moment provide data at up to 1 Mbits/s but this can be expected to increase in the future.
- *Data-only solutions for residential and business premises*. These solutions are currently used for applications such as Internet and LAN interconnection.
- *High-speed and very high-speed links*. These are sometimes configured on a point-to-point basis, providing data rates anywhere in the range 10 Mbits/s to 1 Gbits/s. The higher data rate systems use optical or very high frequency solutions.

The area is complex because for each of these different applications there are multiple competing technologies including cable modems, ADSL modems, powerline technology and other wireless technologies; because there are multiple contending fixed wireless technologies; because the business cases are unproven; and because the market is embryonic. A summary of the key issues is provided in this section.

### 4.3.2  Key Drivers for Fixed Wireless

Fixed wireless has historically been considered as a means of providing fixed connections into homes and more recently as a means of local loop competition. This historic paradigm has had mixed results. More recently, this has changed to an assumption that the role of fixed wireless is associated with the delivery of high-speed data. Most homes that are interested in high-speed data typically already have some form of connection such as copper cabling. In this case, fixed wireless is a competitor, potentially offering better service, lower cost, or both.

### 4.3.3 Key Competitors to Fixed Wireless

There are a range of different possible delivery mechanisms for high-speed access to the premises. These include twisted-pair cabling, coaxial cable (often referred to as 'coax'), fibre optic cable and power-line cable. The key mechanisms are discussed briefly below.

*The copper infrastructure*, comprising twisted-pair copper cables and installed and owned by the state telephone company (the PTO), has formed the traditional access infrastructure for almost 100 years. For much of its life it was ill-suited to anything other than voice due to its low bandwidth (only around 3 kHz). However, the deployment of DSL in the 1990s changed this dramatically. This is currently delivering around 2 Mbits/s on the downlink and penetration levels are approaching some 30% in the developed world. Future advances with ADSL2+ promise data rates up to 20 Mbits/s although, as with all DSL solutions, data rates decline with increasing distance from the local exchange. Some users as a result are unlikely to achieve data rates above 1 Mbits/s unless there is a substantial infrastructure upgrade.

*Cable* operators have implemented buried networks, which initially provided television distribution, in numerous countries. As a result, cable tends to be deployed in residential areas but not business districts. Cable networks vary in their composition. Some are entirely coax. Others use fibre optic cable in the backbone but coax in the local loop; these are often known as fibre to the curb (FTTC) or hybrid fibre coax (HFC).

This is not quite the whole story. For each premise there is one (or two) twisted copper pairs running from the switch right to the premise. In a cable network, all premises share the local branch to which they are connected. To put it differently, all premises on one branch are connected to the same cable whereas they are all connected to their own individual twisted pair. This is fine whilst cable is delivering broadcast services, to be watched by many simultaneously, allowing 50 or more TV channels. However, if each user on a branch wanted a video on demand service, then only 50 users could be accommodated on one branch, unlike twisted pair using DSL where as many users as homes could be accommodated. Cable operators are currently 'shortening' branches by running the backbone further out towards the subscribers.

Cable modems are in widespread commercial service with a downstream capability of 30 Mbits/s and an upstream capability of 10 Mbits/s. Clearly, where cable networks have already been installed, cable operators are well placed to provide a convergent offering with voice, computing and television all through the same channel.

The 'ultimate' local loop might be *fibre* direct to the home. Depending on the equipment used at each end, fibre has a virtually unlimited bandwidth (in the order of gigabits/s). Fibre deployment is expensive, costing around $1500 per home on average because of the need to dig trenches and lay cable. However, in 2006, there were a number of deployments taking place globally in Korea, Japan, the USA and Italy. As data rate requirements to the home rise, it might be expected that fibre penetration will also increase.

### 4.3.4 Likely Success of Fixed Wireless

Key to fixed wireless being successful is its ability to provide a competitive offering to copper and in some cases cable and fibre. This has always been a problem for fixed wireless. Copper technologies have tended to advance at least as fast as wireless, allowing data rates over copper to exceed those over wireless. Prices for services such as ADSL have continued to fall, making it progressively more difficult for fixed wireless offerings to succeed. Competing with cable and fibre optic is even more difficult because of their ability to offer the 'triple play' of video, voice and data. This inability to provide a competitive offer has resulted in virtually all fixed wireless networks failing over time. However, it has not stopped operators continuing to try. Most recently, operators are planning deployments based on WiMax technology. The advantages of this are claimed to be its standardisation, increasing competition and economies of scale and thus reducing equipment prices. It is also claimed that indoor installation is possible, removing the need for expensive outdoor antenna installations.

Our view is that fixed wireless, as a purely fixed offering (see below), is unlikely to succeed beyond a niche of users. Copper offers clear advantages in terms of cost, data rates and overall capacity. Fixed wireless systems provide a shared data pipe and as more users in a cell become subscribers, the cell capacity will become increasingly stretched. Standardisation is unlikely to provide a sufficient cost reduction to offset the ever falling cost of ADSL provision. Indoor operation will inevitably reduce cell size as the building penetration loss reduces the signal strength, further increasing the network deployment cost. It will also make the service less reliable as it is possible that the antenna will be moved to areas of poorer coverage in the home or placed near sources of interference such as domestic appliances. Fixed wireless has not succeeded in the past and we do not see any fundamental change in the factors that have brought this about.

## 4.3.5 Enlarging the Market with a Nomadic Offering

One option considered by some fixed wireless operators is to deploy a combined fixed and nomadic network. Here subscribers can access the service from within their homes as a normal fixed access system, but also can use the service when out of the home, either when close to the home, or in other coverage areas such as city centres. The service is termed nomadic rather than mobile to indicate that the user may need to locate an area of coverage and then remain broadly stationary while accessing the system.

This does provide a further feature that may be some value to the user and cannot be replicated by other fixed services. Hence, it will likely increase the chances of success somewhat. However, our view is that the user will see these benefits as only slightly valuable and will not be prepared to pay a substantially increased monthly fee. Hence, it is unlikely to significantly change our conclusions as to the viability of fixed wireless.

## 4.3.6 The Prognosis for Fixed Wireless

Since its inception in the late 1990s, fixed wireless has failed to become successful. Although the market is large enough to support a few medium-sized companies it has remained small compared to other wireless solutions like cellular and W-LANs. At present it is difficult to see what might change this situation. Although demand is growing for high-speed connections, this demand is generally well served by wired solutions. The possibility of a breakthrough technical solution seems remote, as discussed above and in more detail in Chapter 6. Perhaps in the longer term, if data rate requirements to the home rise above the levels that can be supported by ADSL, if fibre is not deployed more deeply into the network, and if technologies such as mesh have become established, then fixed wireless might become viable. But for now it is hard to see it playing a key role in the future of wireless technology.

## 4.4 Short-range Devices

### 4.4.1 Introduction

An area of proliferating standards and technologies in the period 2000 to 2005 has been the world of short-range devices. It has also been one of the few areas of wireless communications reporting rapid growth. Shipments of BlueTooth chipsets and W-LAN devices are growing rapidly year-on-year and are soon likely to reach a stage where most handsets have BlueTooth imbedded within them, and most laptops have W-LAN capabilities.

In order to understand this market it is first worth understanding why anyone would want a short-range device. In practice, few people want a device specifically because it has a short range – for most users the longer the range the better. Devices are designed to be short-range for one or more of the following reasons:

- A high data rate is required which implies either smaller cells or the higher frequency bands, which in turn have reduced propagation. Small cells can also have simpler access protocols as less interference and propagation delays need to be taken into account, speeding the data rates.
- Long battery life is required which results in low power transmission.
- Low cost is required which results in low-power RF components integrated into the chipset.

Different applications make different trade-offs on each of these axes – for example some require the highest data rates, others the lowest cost, others a certain mix of each attribute. This explains the proliferation of different standards covered in more detail below. Potential applications for short-range devices are those that are not well suited to cellular for one or more of the following reasons:

- Cellular cannot provide a high enough data rate.
- Cellular cannot provide the functionality needed (such as ad-hoc networking).
- Cellular call charges would be too expensive for the application.
- Cellular technology is too costly to be integrated into the device.

Falling into one or more of these categories are the following types of application:

- *Networking around the office or home.* Here cellular is too slow and too expensive. In future, this area might also include networks in aircraft and trains, as well as personal area networks around the body.
- *High-speed data transfer.* Here cellular is too slow. Specific applications include email synchronisation on the move in hotspots and video transmission from a camera to a display device.
- *Cable replacement applications.* Here cellular is too expensive and the technology too costly. There are many well-known applications here such as wireless headsets, wireless mice, car hands-free kits and PDA to computer connectivity.
- *Machine-to-machine communications.* Here cellular is too expensive and complex; however, if large distances are involved it may be the only option. Specific applications in this area include remote sensing (e.g. of growing conditions in a field), and the replacement of physical keys for security purposes.

Although short-range devices fill in those areas where cellular is inappropriate, in some cases it is useful to integrate the two together, for example using the cellular handset within the home via short-range technology, or the 'master' in a machine-to-machine subnet being able to signal back to the control room via cellular. Generally, this integration is mostly a matter of higher layer software rather than requiring standards to be integrated.

While there are relatively few uses for short-range devices at present, there is a strong potential for the 'tipping point' to make short-range applications widespread. The tipping point is a description for a process whereby once penetration exceeds a certain point a virtuous circle sets in which rapidly drives growth. For example, if a manufacturer of training shoes perceived that a sizable percentage of their target market also had home computers fitted with BlueTooth they might decide to market a wireless training shoe. Such a shoe might capture data when out running – such as number of strides, weight distribution etc. – which a keen runner could download to a computer for analysis of the run. If a number of such applications became available this would encourage more people to get BlueTooth for their computer, which would in turn stimulate more applications, and so on. The trick is getting penetration past the tipping point where market forces can drive it rapidly upwards. Given the relatively low cost of short-range devices and the widespread belief in them by companies like Sony, there is every chance this might occur. The biggest obstacle is the number of short-range standards – only if most users have devices conforming to the same standard will the tipping point be reached.

In the next section we look at the different standards that are emerging and the reasons why the market has become so fragmented.

### 4.4.2 Overview of the Standards for Short-range Devices

The different short range technologies either available or currently under development are shown in Table 4.1, ordered approximately by their maximum data rate.

**Table 4.1**   Short-range technologies.

| Standard | User data rate | Range | Approximate price in 2006 | Battery life |
|---|---|---|---|---|
| UWB | >100 Mbits/s | 5–10 m | $5–10 | Hours to days |
| W-LAN 802.11 | 5–30 Mbits/s | 100 m | $20 | Hours |
| BlueTooth | 700 kbits/s | 10 m | $3 | Hours |
| DECT | 500 kbits/s | 100 m | $5 | Days |
| Zigbee | 128 kbits/s | 30 m | $1 | Weeks to months |
| RFID (incl. NFC) | 1–200 kbits/s | 0.1–10 m | $0.05–1 | Weeks to indefinitely |

A more detailed description of each of these technologies is provided in the section below, but in outline:

- Ultra Wideband (UWB) uses a novel modulation format to transmit across an extremely wide band. Subject to regulatory approval this enables it to operate in spectrum already in use. Its key advantage is the potential for extremely high data rates but the need to operate at low power results in a short range for many applications.
- The W-LAN standards, including most importantly 802.11a, b and g, are the de-facto W-LAN technologies, with 802.11b already in widespread deployment. They provide a high data rate and a relatively high range but at the expense of cost and battery life.
- DECT is well known as the cordless voice technology used throughout Europe. It can also carry data. Its advantages are range and price, against which it trades off a relatively low data rate and a battery life that is average.
- BlueTooth is also well known as the short-range standard. Its key attribute is low price against which it trades relatively low data rates, range and battery life. It is becoming increasingly widely embedded in many devices.
- Zigbee is an emerging standard that is intended to replace BlueTooth in applications that require much longer battery life than BlueTooth can support. To do this it has a lower data rate as well as a lower duty cycle for transmission and reception.
- Radio frequency identification (RFID) tags form a wide-ranging category from passive devices printed with special ink on paper labels through to active transponders. It includes a standard currently being developed called 'near-field communications' (NFC). Tags are typically low cost and can have extremely long battery life (or indeed, not need a battery at all) but offer very low data rate and a range that is often just a few centimetres.

The advantage of having such a wide range of different technologies is that the designer of a specific application can choose one that is likely to be well-tailored to their particular needs. The disadvantage is that there is little guarantee that a user will have the same technology embedded in the device that will form the other side of the link. The key to determining which technologies will succeed is in understanding this balance between optimised functionality and widespread penetration. The final part of this section takes a look at this balance and makes some predictions about the future for short-range devices.

### 4.4.3 Ultra Wideband (UWB)

The idea of a W-PAN technology emerged during 2002 within the IEEE 802.15.3 group and has been named WiMedia. This technology is broadly targeted at the distribution of audio-visual information around the home. So, for example, it might be used to wirelessly connect a DVD player with a screen and speakers, potentially not in the same room. To do so, it needs to be able to provide data rates of at least 20 Mbits/s with a high quality of service. This is not something readily achieved with the current W-LAN standards which do not offer reserved timeslots and hence the apparent need for a new standard.

Initially the WiMedia forum looked at a range of different air interface standards with different data rates. In 2004, they decided to concentrate on UWB as their key technology. UWB provides data rates between 100 Mbits/s and 1 Gbits/s but this may be at the expense of reduced range, potentially limiting it to within a single room. Whether this increased data rate but reduced range would be valuable to users is not currently clear.

In 2006, WiMedia was still a developing technology, with chipsets and equipment on the verge of appearing. Its success will depend on its adoption by manufacturers of home audio-visual equipment.

## 4.4.4 Wireless LANs [5]

Almost all W-LAN solutions conform to the IEEE 802.11 standard. At the time of the first version of this book, the HiperLAN2 standard was also a contender, but since 2000 support has mostly evaporated. Hence, in the remainder of this section we concentrate on 802.11 and its many variants.

An 802.11 W-LAN system is typically comprised of one or more base stations or access points and a number of mobile devices. W-LAN devices can operate without a base station in the ad-hoc fashion that BlueTooth utilises (see below) but this mode of operation is far less common. The IEEE 802.11 protocol stack consists of the lower physical layers and then a common layer-three protocol.

There are many different variants or amendments of 802.11 – at the time of writing the key ones were:

- 11a: a high-rate version of 802.11 operating in the 5 GHz band
- 11b: now considered the standard W-LAN solution, operating at 2.4 GHz
- 11g: a high-speed variant but operating in the 2.4 GHz band.

They all share the same multiple access method – carrier sense multiple access with collision avoidance (CSMA/CA).

W-LANs suffer from the same hidden terminal problem as cognitive radio, and this approach attempts to resolve the problem, albeit with some interference between users. Essentially, a device that wishes to transmit a packet monitors the radio channel. If it detects activity it does not transmit. If it senses no activity it waits a random time before sending its packet. It is possible that another device also transmitted its packet at the same time. To overcome this, the base station sends an acknowledgement for each packet received. If the transmitting station does not get an acknowledgement it waits a random length of time before re-transmitting its packet.

The technologies differ in the physical layer they adopt. A summary of the similarities and differences between the standards is shown in Table 4.2.

**Table 4.2**  A comparison between different variants of 802.11.

| Characteristic | 802.11a | 802.11b | 802.11g |
|---|---|---|---|
| Spectrum | 5 GHz | 2.4 GHz | 2.4 GHz |
| Max physical data rate | 54 Mbits/s | 11 Mbits/s | 54 Mbits/s |
| Max user data rate | 25 Mbits/s | 5 Mbits/s | 25 Mbits/s |
| Physical layer | OFDM | DSSS | OFDM |

On the face of it, there would seem to be a significant problem of proliferating W-LAN standards, leading to end user confusion and eventually lack of market success. In practice, this has not happened because the 11b standard has had some time to become the established de-facto standard. At present, if manufacturers choose to use 11a or 11g they add it to a product in addition to 11b. This provides a level of stability in the marketplace and ensures that all products can interoperate and can make use of higher speed versions if they are available at both the base station and the mobile.

One of the key developments in wireless communications since 2000 has been the success of W-LANs. In 2000 it was not clear how widely they would be adopted, but by 2006 virtually all laptops and many PDAs had in-built W-LAN capabilities. W-LAN base stations for home and offices were also growing at an ever increasing rate. Hence, the widespread success of W-LANs in the home and office seemed assured.

Less clear remained the hotspot market where W-LANs are deployed in locations such as airports to provide a public high data rate service. This is a market that had experienced problems in 2000 in the USA where both major players – Metricom and MobileStar – went bankrupt. By 2006 the market appeared to have recovered with a number of the major cellular operators such as T-Mobile and Swisscom making significant investments in W-LAN hotspots and entities like The Cloud continuing widespread deployments. However, even in 2006 it was unclear whether public W-LANs would be successful and what the on-going business model would be.

Our view is that public W-LANs are a very good match to the needs of laptop users. With W-LAN capability built into laptops no further devices are needed. W-LANs provide a fast and effective way to transfer emails and browse the Internet and have already been implemented in most key travel locations. This seems a viable, but not particularly large market. However, if integrated with other services as part of a convergence process they become of significant value. We expect W-LAN hotspots to be increasingly embraced by major cellular and fixed operators around the world.

## 4.4.5 BlueTooth [6]

An individual may have a number of devices that he or she carries on or near the person. The concept of cable replacement technologies like BlueTooth is that if each of these devices had in-built short-range wireless communications they could exchange information without wires and without intervention from the user. So, for example, a laptop, which was ostensibly in 'sleep mode', stored in a briefcase, could periodically talk to the cellphone clipped to the user's belt and ask it to check for emails. The cellular phone could retrieve these and send them to the laptop over the short-range link. The laptop could then store them so that, when the user turned the computer 'on', all the emails would be available on the computer. As the user performed actions such as sending emails, the computer would talk with the mobile phone and request transmission of these to the data network.

This kind of concept results in a number of unusual requirements and properties, often talked about under the term 'ad-hoc networking'.

- The short-range devices must be very inexpensive as they need to be embedded in most machines. Price points of a maximum of $5 per chip are often discussed.
- This price requirement in itself tends to limit the devices to very short range since one of the major cost elements of a wireless device is the RF amplification.
- The system needs to utilise unlicensed spectrum and needs to be 'self-organising' from a spectral viewpoint as there can be no central control.
- Networks need to be 'self-discovering' and 'self-managing'. No one device can know beforehand what other devices might be in the vicinity, so devices need to have a method for discovering other devices, finding out what they are and exchanging appropriate information with them.

- With an almost infinite range of devices, understanding how devices communicate, what information they might want to send to each other and whether they have appropriate rights to do so becomes problematic.

BlueTooth is designed to operate in the unlicensed industrial, scientific and medical (ISM) band centred around 2.45 GHz. The bandwidth of transmission is 1 MHz, but the actual 1 MHz band utilised within the 80-MHz wide ISM band can be selected by the device depending on the environment. In order to overcome the anticipated interference within the ISM band, BlueTooth utilises frequency hopping such that a single (static) source of interference will only have a limited effect on transmission quality.

Two levels of transmit power are available within the BlueTooth specification, 1 mW (0 dBm) and 100 mW (20 dBm). The former allows transmission ranges of up to around 10 m, the latter up to around 100 m. There is still some debate over which of these power levels is more suitable. The lower power levels result in inexpensive devices and minimal interference to other neighbouring PANs. The higher powers are more expensive to implement and have a greater drain on the battery, but enable simpler coverage of a house or other building. The higher power devices utilise power control to enable them to communicate with lower power devices and to reduce transmission power where full power is not required.

One of the largest problems for BlueTooth is detecting the presence of other users and building an ad-hoc network. This is exacerbated by the general need for BlueTooth equipped devices to have low battery drain. The only way for one BlueTooth device to discover another is for at least one of them to transmit periodically, however, the frequency of transmission is kept low to minimise battery usage. Since the device will transmit for only a short time period and may do so on any one of the numerous different frequencies in use, the receiver must listen as long as possible. However, this also drains batteries. Complex algorithms are in place to enable devices to become aware of each other with a minimal battery drain but within a sufficiently short time period that the user is not waiting for the devices.

After some false starts, the BlueTooth market has entered a period of solid growth. BlueTooth is now included as standard in many phones, laptops, PDAs and computers. BlueTooth headsets have proved to be particularly successful.

## 4.4.6 DECT

DECT is a well established standard for cordless telephony. DECT phones and base stations are now widely installed in homes and offices. DECT is a relatively old technology, using low compression voice coders and simple modulation in the 1880–1900 MHz band.

DECT is also able to handle data up to rates of around 500 kbits/s. However, to date it has had limited success in this role because of the higher speeds offered by W-LANs. Looking forward it seems likely that DECT will continue to be the cordless voice technology of choice for many years, eventually overtaken by whatever solution transpires in the in-building space.

## 4.4.7 Zigbee

The rationale for Zigbee arose from a design choice made in BlueTooth. Because of the manner in which BlueTooth has been designed it uses batteries relatively quickly – devices like BlueTooth headsets may need to be charged daily. There is a wide range of applications where longer battery life is desirable – for example in a cordless mouse – while the high data rates and advanced capabilities of BlueTooth are not needed. This is what Zigbee has been designed for. The key parameters for Zigbee are:

- data rates in the range 10–115 kbits/s
- 30 m range
- up to 255 nodes in a network (compared to eight with BlueTooth)
- 0.5- to 2-year battery life
- 5 MHz channel spacing.

In 2006, Zigbee had completed standardisation but devices had not yet emerged in quantity. BlueTooth manufacturers were also 'fighting back' with very low power versions of BlueTooth. At this time it remains difficult to predict whether Zigbee will be successful.

### 4.4.8 RFIDs

This is a broad category of devices used mostly for tagging purposes. RFIDs are routinely used in shops to prevent theft, in baggage systems to track suitcases and so on. Increasingly, they are also being used in smart payment systems where a user can hold a pass near a barrier in order to open it.

RFIDs divide into passive and active systems. Most solutions are passive. When a passive tag is in the presence of an electrical field it radiates back some of the energy, modified in accordance with the information held on the tag. This allows passive tags to be cheap, not to need battery replacement and to be very small. However, it limits the range to a maximum of around 30 cm, sometimes less. Active solutions listen for an interrogating signal and transmit back the appropriate information. They can have a range of many meters, and indeed become more like BlueTooth or Zigbee devices.

Although a relatively large market, RFIDs are somewhat at the periphery of mainstream wireless communications and will not be considered further here.

### 4.4.9 The Prognosis for Short-range Devices

The history of standards tells us that generally only a few survive, particularly where interoperability is important. Of the three competing standards for video cassettes, we have been left with VHS – the least technically advanced but critically the first one past the tipping point. In the world of audio, CDs became successful but the digital compact cassette and minidisc have been consigned to also-rans. In the world of cellular the standards are slowly being whittled away, with the US TDMA standard now fading, leaving only GSM and CDMA. It is hard to think of an example of an area where some six standards, the current number of short-range technologies, all survived. History also shows us that where one standard has become adopted before another is launched it is very hard to displace unless the new one is overwhelmingly superior. Indeed, some have suggested as a rule of thumb that the new technology needs to be 'ten times better' than the old if it is to persuade almost all users to adopt it. Conversely, where a number of competing standards are all launched simultaneously it results in market confusion and often none succeeds.

In the world of short-range devices there are two standards that have been launched and are well on the way to success. The 802.11 W-LAN standard is now being embedded in over half of all laptops made. Equally, BlueTooth is embedded in all high end cellular handsets and a number of digital devices like PDAs. DECT is in widespread use as a voice technology in the home and office, but has achieved little in the way of data usage. With such a large lead, other technologies can only catch up if they are much better, or sufficiently different from W-LANs and BlueTooth, or if flaws are discovered with these leading technologies.

Some technologies can be relatively easily dismissed. DECT is interesting because of its ability to carry both voice and data, but its data rates are too slow to make use of this capability and it seems likely it will remain a voice technology. UWB is very different from other technologies, but it is unclear whether there are sufficient applications that need its very high data rates and W-LANs are so much better in many other aspects. (Although it is worth noting that UWB has other important applications such as short-range radars that are not considered here.). That leaves us with Zigbee and RFID tags as more serious contenders. Before we examine them in more detail we need to first look at potential flaws in the two current 'winners'.

The 802.11 W-LAN standard seems to have few major flaws. There are a number of issues surrounding security, quality of service and so on, but these are being fixed as part of updates to the standard. W-LANs are now sufficiently widespread that finding a major flaw seems unlikely. BlueTooth is steadily improving after a period where many devices were not interoperable and the effort needed to pair devices was sometimes more than many IT-professionals were prepared to put up with.

Zigbee has a key advantage over all other similar short-range standards in terms of its battery life. This will be particularly important in some applications, often those not requiring interoperability with existing devices. For example, in the deployment of Zigbee devices throughout a farmer's field to sense growing conditions it matters little what standard the farmer has in his home computer. Zigbee is likely to succeed as a niche standard for specific applications where widespread interoperability is not needed.

That leaves RFID tags. These are generally used in quite different applications from other short-range devices such as baggage labelling. Like Zigbee, there is little reason why they cannot continue to be successful here.

So, in summary, W-LANs and BlueTooth will dominate the short-range devices market. Although not optimal for some applications, the widespread deployment of these technologies in a range of devices will spur many unforeseen applications along the lines of the wireless training shoe and self-setting alarm clock. The resulting virtuous circle will deliver rapid growth in the short-range devices market over the next decade, transforming devices and the way we interact with them.

## 4.5 Core Networks

A short discussion of networks is worthwhile, since almost all communications make use of networks to form the route between the originating and terminating parties. At the heart of a network are a number of switches, which form routes for calls across the network. Around the switch there may be many other components such as location registers, databases and subsidiary switches. All of these are connected via broadband, typically dedicated interconnections. A core network, for example for a cellular system, can be a highly complicated entity.

Core networks are rarely constraining in terms of capacity. It is much simpler and cheaper to add more capacity to the core than it is to the radio network. However, core networks can be highly constraining in terms of services and in the way that convergence is achieved. It is this factor that we will explore here.

Historically, networks have been integrated with particular access technologies. So a cellular network has its own core network, as does a fixed access system. The cores are somewhat different, having different databases and slight protocol variations, but typically share similar platforms and technologies. This vertical integration of core network with access network has arisen because:

- often the owners of the different networks have been different entities and they have been unwilling to share core networks
- each system has required a slightly different core network and it proved simpler and cheaper to construct separate networks then attempt to deploy a more flexible general purpose core.

Equally, vertical integration has not to date been seen as much of a problem. Convergence has been limited and the cost of the core network is typically less than 10% of the total network so the disadvantages have been seen as minimal. Going forwards, we predict that convergence will increase pressure on the status quo.

The other disadvantage of existing core networks is in the simple introduction of new services. Services such as call redirection and three-way calling are currently provided by the core network, typically on the switch itself. Adding a new service requires an upgrade to the switch. This can often be done only by the switch manufacturer and is seen as a major event with the potential to disrupt the whole network. This tends to make it slow and expensive. Many efforts have been made to overcome this including a standardisation of the 'intelligent network' concept. Here, the software needed to describe a service is separated from the switch, in principle allowing other entities to provide the software. However, for various commercial reasons 'intelligent networks' were implemented to only a limited extent.

The drivers for change came, interestingly, not from within the community working on core networks, but from the Internet. The Internet was, from the start, based on packet transmission whereby information is encapsulated into short packets, each one given a destination tag, and sent into the network. This contrasts with most telecommunications core networks which use circuit switching whereby a dedicated path is established between originating and terminating parties along which information can flow for the duration of the call. The advantages and disadvantages of each are well rehearsed. Broadly, packet switching was long thought unsuitable for 'real time' communications – calls such as voice or video where the variable delays associated with packet transmission could be annoying to the user. Indeed, this is still true to some extent today, with VoIP calls occasionally having a lower quality than conventional calls.

Packet switching networks can have a simpler core. All that is needed is a set of routers, able to examine the packet headers and forward appropriately. This simplicity lends itself well to having a different 'service plane' where instructions are stored for more complex services such as voice – effectively an identical principle to intelligent networks.

The growth of the Internet, and the flexibility of Internet Protocol (IP) based networks, has, over time, persuaded many that the future of core networks is to move from the existing vertically integrated cores towards common IP networks. In this vision, a few generic IP cores are connected to multiple different access networks. Each access network encapsulates its users' information in IP format and sends it on to the core. Specialist functions such as mobility management for mobile networks are provided by additional servers sitting adjacent to the core network. The advantages are:

- Convergence is much more readily achieved when access networks share a common core, or when all cores are similar, since, for example, common services can be provided by the same server across different access networks.
- Cost is expected to be lower since routers cost less than equivalent switches and IP is typically a more efficient use of resources than circuit switching.
- Flexibility is greater since services can be simply introduced.
- Multi-media services are much simpler to handle. For example, if a user decides to change from voice to video during a call, a circuit switched network would need to establish a new connection type. A packet network simply generates more packets.

The disadvantages relate to the concerns over the ability of IP networks to provide low-delay, low-latency calls for real-time communications. These concerns are easing as modifications are made to the IP protocols; for example, IP version 6 (IPv6) or multi-protocol label switching (MPLS) both enable quality of service (QoS) to be specified and traffic treated accordingly within the network. Many mobile operators now expect to deploy what is termed an IP Multimedia System (IMS) – the idea of a core network based around IP, able to handle a wide range of different traffic types, and with intelligence provided by a range of servers situated around the edge of the core network. For voice, the service is often based on a protocol called Session Initiation Protocol (SIP) which provides the necessary intelligence to establish the route for IP packets at the start of the call.

The increasing prevalence of the Internet, the advantages of IP core networks, and the gradual advances overcoming the problems of packet switching have slowly, over time, convinced almost all that IP-based core networks will become standard in the future. The remaining difficultly is in understanding the timing.

The all-IP vision has now been commonly held for some 5–10 years. So far, apparently little progress has been made against that vision. Instead of a large switchover from circuit switched to packet switched systems, operators have opted to gradually introduce packet switching alongside their circuit switch. However, a number of operators, such as BT, have announced major switchover plans – in the case of BT to their '21st century network'. This is expected to be achieved by 2010. BT are one of the leading operators in this space; hence, it will be 5–10 years before the all-IP vision starts to get realised.

## 4.6 Broadcasting

### 4.6.1 Conventional Broadcasting

Broadcasting broadly covers audio and video broadcasting – both can be treated in the same way for the purposes of this discussion. Considering, say, video broadcasting, there are the following mechanisms:

- *Terrestrial broadcasting* using UHF channels to broadcast video signals. This remains the predominant means of broadcasting in many parts of the world.
- *Cable broadcasting*, sending multiple channels down a cable. Since this is not wireless, it is not a key focus for this book.
- *Satellite broadcasting* using a geo-stationary satellite to broadcast multiple channels to receiver dishes mounted on the side of homes. This is a mechanism currently gaining ground due to the number of channels that can be distributed and the relatively rapid manner in which it can be established compared to other alternatives.
- *Internet broadcasting* is still in its infancy but some are now developing streaming media systems whereby if end users log on to a particular site they will receive whatever 'station' is currently being broadcast. Although the end user would perceive this to be broadcast, from an architectural point of view, since the signal is sent individually to each subscriber, it is more of a multiple one-to-one communications mechanism than a true broadcast.

Over the last few years there has been a slow move towards 'personalisation' in broadcasting. This is the concept that instead of watching whatever is broadcast on an available channels, that a 'personal' channel is assembled for each user based on his or her preferences. The first moves to do this were the 'TiVo' home personal video recorder (PVR) around 2000. PVRs have slowly grown in popularity and now are reasonably successful in understanding a user's preferences and recording appropriate content on to their hard disk. The next stage in this process might be for home media centres to seek content from the Internet to supplement the broadcast content in delivering personal TV channels, although this is a substantially more complex problem than selecting from broadcast content. If this approach were to become ubiquitous then broadcasting, as an architectural way to reach multiple subscribers simultaneously, would slowly cease to exist. Home media centres would increasingly rely on downloading material from the Internet.

The argument for personalised broadcasting is somewhat weakened by the cultural change that would be required to bring it about. People enjoy gathering together to watch sporting events, or tuning in every night to watch their favourite soap opera. Cultural change of this sort takes at least a generation, and so although we might see significant change in the next 20 years, we are unlikely to see the death of broadcasting as we currently know it.

High-definition transmission currently requires around 5 Mbits/s, standard definition around 1.5 Mbits/s and low definition suitable for mobile screens around 200 kbits/s. These figures will reduce over time as compression technology improves. Providing 1.5 Mbits/s to the home using fixed networks is perfectly feasible and 5 Mbits/s can be expected within a few years. Providing 200 kbits/s over cellular networks is also viable, although only to a subset of subscribers before the total capacity of the cell is reached.

So, in summary, it would be technically possible and attractive to many to move to a world where broadcasting is personalised. However, equally it requires significant cultural change of a form that is likely to take a generation. And whatever happens, people will still want to tune in live to a number of key events.

There are many interim stages between the current situation and full personalisation. For example, early 3G launches offered highlights from key football matches. These highlights are collected by the broadcaster, edited for transmission over a cellular network and then passed on to the cellular operator for them to 'broadcast'[5] to those who have subscribed to the service.

Another important area is audience participation. Many programmes now allow the viewers to vote, perhaps for their favourite celebrity. There are two key mechanisms for doing this – via the fixed line or via the mobile phone. To date, the mobile phone, using SMS, has proved to be the most popular mechanism. This works well, but could be better integrated so that, for example, a list of participants appears on the phone and the viewer just has to select one and press OK, rather than composing an SMS.

## 4.6.2 Mobile Broadcasting

Around 2005, the concept of mobile broadcasting was much discussed and a number of trials took place. Rather than individually sending video streams over 3G, which as discussed above rapidly becomes expensive, the mobile TV systems broadcast the same content over a wide area. Of course, mobile TV receivers are not new – it has been feasible for around 20 years to watch TV on a small mobile device. The differences with the proposed mobile TV system are:

- It can readily be integrated into a mobile phone.
- Through using digital transmission with coverage tailored to mobile devices the availability of the signal and the quality of reception can be higher.
- By careful design the battery load on the mobile device is lower.
- Through using digital compression, more channels can be made available than with conventional analogue TV.
- Error correction and interleaving is included in transmission to make the reception more resilient to fading and other effects of movement while viewing TV.

There are many competing standards for mobile TV, including digital video broadcasting for handhelds (DVB-H), digital multimedia broadcasting (DMB) and proprietary solutions like MediaFlo.

Predicting whether mobile broadcasting will be successful is difficult. On the one hand, few have shown an inclination to watch broadcasting when mobile using existing analogue receivers. On the other, the advantages listed above are substantial, especially allowing viewing on a mobile device that most carry regularly with them. New programming may be needed as research suggests that users want to watch very short 'mobisodes' perhaps lasting just a minute or two, to fill in quiet periods in the day. But whether users will pay for this content is unclear – especially given that in many countries much of the broadcasting content is provided free. Users can also download 'video podcasts' from the Internet when at home for storage on their devices for later viewing, which may prove more cost-effective than developing a mobile broadcasting network. Overall, we remain slightly sceptical of the need for separate networks for broadcasting to mobile devices, but there may be many convergent models with cellular transmission that enable mobile TV in innovative manners.

---

[5] They are actually sent as multiple individual messages rather than broadcast in a one-to-many format. This results in some inefficiency if there are multiple users in a cell, but is simplest to implement.

## 4.7 Industry Structure

It is worth discussing the structure of the communications industry that was in existence in 2005. This is because, as we will see in later chapters, this structure might be an impediment to future advances in technology and services.

The communications industry has developed predominantly in specialised 'verticals'. For each type of service, such as cellular, or fixed, there is one or more companies, who have built the network, run the network, deliver the customer service and billing, maintain a brand, etc. These companies might include:

- A fixed operator (typically the old 'post and telecommunications' entity). In the USA the single provider was split into multiple regional providers, who are now recombining in various ways.
- One or more cable network operators.
- One or more terrestrial TV broadcasters.
- Typically a single satellite broadcaster.
- Multiple cellular network operators.
- Multiple W-LAN hotspot providers.

There have been some attempts to change this. Mobile virtual network operators (MVNOs) act as service providers on top of 'network pipes'. Some operators are outsourcing the building and maintenance of their networks so they can concentrate on service provision, taking the first steps towards disaggregation. Some cellular operators have also become hotspot providers, and third parties have emerged who build hotspot networks and then sell capacity to service providers such as cellular operators.

Nevertheless, the industry structure is still predominantly one of vertical silos with different communications technologies competing with each other at the edges. Such a structure makes the provision of consolidated services difficult. If network provision was separated into 'pipes' and 'services' then a service provider could buy bulk capacity from a range of network operators and offer a consolidated service to end users. Instead, consolidated services can only be offered if operators reach agreement between themselves. This is often difficult to do and can restrict the offering to a subset of available operators.

It seems unlikely that change will happen soon. Companies rarely choose to split themselves apart. Even under regulatory pressure, BT recently reorganised along a pipe/services division but did not go as far as creating separate companies. This could be a significant impediment to the provision of consolidated services in the future.

## 4.8 Summary

Most of this chapter has been devoted to ensuring a good understanding of the current position. Key observations were:

- 3G will be increasingly deployed, becoming the key cellular system within the next five years. 2G will continue to exist in parallel for at least another decade. Cellular operators will enter a long period of stability.
- WiMax is unlikely to make a major impact on the success of 3G.

- There may not be a discrete 4G, based on a new standardised technology; instead there will be incremental improvements in existing standards and their interoperability.
- Convergence between cellular and wired networks will likely be based on W-LAN in the home and office and cellular in the wide area.
- Mobile mesh systems do not have sufficient advantages to bring about their widescale deployment but may have many niche applications.
- Fixed wireless will likely not succeed substantially, even with the introduction of a nomadic offering.
- The key short-range devices will be W-LANs providing building networks and Bluetooth providing device-to-device connectivity.
- Core networks will migrate to IP over the next decade.
- Conventional broadcasting will continue long into the future.
- Building a new network for mobile broadcasting seems hard to justify, although it is difficult to predict demand.
- The current industry structure is for providers to be vertically integrated. This will slow the speed of convergence due to the need to form partnerships rather than just procuring services.

## 4.9  Appendix: The Role for OFDM

### OFDM is Increasingly in Favour

Just like clothing, wireless access methods seem to move in and out of fashion. In the mid 1990s, CDMA was the fashionable access method, being rapidly adopted by the designers of cellular, fixed wireless and W-LAN solutions. In the early 2000s, OFDM seems to have taken on the mantle of the 'up and coming' access solution. Already used in digital sound and TV broadcasting, OFDM is being deployed in W-LANs and fixed wireless systems and is being proposed by many as the likely technology behind 4G.

While the comparisons with fashion may seem a little far-fetched, it is likely that design engineers are influenced by the systems emerging around them and also that strategists and marketers see advantages in offering a technology that seems to be in favour.

There were good reasons for the popularity of CDMA – it genuinely provided an increased capacity compared to existing solutions. But are there good reasons for OFDM and is CDMA no longer appropriate? In order to answer these questions we first provide a short overview of OFDM compared to other access technologies so we can demonstrate the advantages and disadvantages it has. We then look at why OFDM is appropriate for some specific applications but not as a general approach to all systems.

### A Quick Introduction to OFDM

OFDM is a complex modulation method, meriting entire books. In outline, with a conventional (often called single-carrier modulation – SCM) scheme the incoming data stream is modulated directly on to a single carrier. With OFDM it is split into a number of parallel data streams each of which is modulated on to a separate carrier. This results in a number of closely spaced carriers, as shown in Figure 4.3, where the carriers can be seen to overlap but because of the orthogonality between them they do not interfere. Each carrier relays a fraction of the user data stream. By making these carriers orthogonal to each other the need for a guard band between them can be avoided and so, in a perfect channel, the bandwidth consumed by both SCM and OFDM solutions is the same.[6] If the transmission channel were perfect, both modulation schemes would provide identical performance, but the OFDM solution would be more complex as a result of the need to generate multiple carriers. However, transmission channels are not perfect, and it is the manner with which both schemes deal with the imperfections that brings about the relative advantages of each.

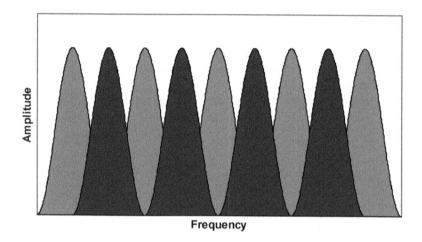

**Figure 4.3**   Representation of OFDM carriers.

### Multipath: the Key Difference between OFDM and SCM

The key imperfection that determines which is the most appropriate scheme is multipath. Multipath results when the same signal reaches the receiver via multiple paths, often as a result of reflections from buildings, hills, etc. Simplistically, if all the significant multipath signals reach the receiver within the duration of a transmitted symbol (e.g. within 1 μs for a symbol rate of 1 MHz), then they result in fading whereby occasionally they add destructively and no signal is received. This can be overcome only by adding redundancy to the transmitted signal in the form of error-correction coding. If the multipath signals are spread over multiple symbols (often known as inter-symbol interference – ISI) then they tend to massively increase the error rate unless an equaliser is used. This device basically collects all the multipath signals, delays them appropriately and then adds them all together. Not only does this solve the ISI problem, it actually enhances the signal quality because even if one signal path becomes obscured the receiver can still receive energy from a different path.

---

[6] Note that some proponents claim that the orthogonality of OFDM provides advantages over other modulation schemes. This is not so, it merely ensures that OFDM is not at a disadvantage.

So, in summary, if multipath is less than a symbol in duration then no equaliser is needed but redundancy has to be added to the transmitted signal. If it exceeds a symbol in duration then an equaliser is needed but the overall signal quality is enhanced. The extent of the multipath depends on the transmitted data rate and the transmission environment.

With OFDM, the symbol rate is reduced by the number of carriers used – a typical example might be 256. This generally makes the symbol period on each of the carriers longer than the multipath propagation time and so no equalisation is needed. However, because the multipath now all arrives within a symbol period it results in fading on some of the carriers and so additional redundancy is needed to overcome this. Understanding this point is critical to understanding the role of OFDM.

### Equalisers may become too Complex to be Realisable

Equalisers are relatively complicated devices requiring a large number of multiply and accumulate operations. The number of calculations per second that an equaliser has to perform rises linearly with both the transmission speed and the number of symbols over which the ISI is spread. As the spread of the ISI is linearly related to the transmission speed the net effect is that equaliser complexity grows as the square of the increase in transmission speed. In many situations this is not problematic – the equaliser in GSM for example is very readily implemented as only a small part of the chipset – but in some systems the complexity can exceed the practical realisable limit of DSP-type devices. This limit is growing as devices become faster but equally so are the transmission data rates attempted.

If the equaliser cannot be implemented cost-effectively then single carrier modulation cannot be used because the resultant interference would destroy the signal. In fact, this reaches to the heart of the role of OFDM. If an equaliser can be implemented then SCM will provide the best performance. If it cannot, then OFDM is probably the only viable solution.

### Problems Specific to OFDM

Before looking at specific applications it is worth noting a number of problems specific to OFDM:

- Redundancy must be added to maintain orthogonality. Any multipath within OFDM reduces the orthogonality and therefore degrades the signal. This is overcome by adding a repetition of the signal on to the beginning of each burst. This can add an overhead of anything between 5% and 20%.
- High-power RF components are needed. OFDM has a high peak to average signal ratio because of the possibility that all channels transmit the same symbol at the same time. To transmit such a waveform without distortion requires a linear amplifier operating well within its limits. Achieving this is typically both expensive and power hungry.

### Specific Applications

Because the impact of multipath is different in each application, it is important to consider the merits of OFDM on an application-by-application basis.

### Broadcasting

There are two important characteristics of broadcast applications that make them quite different from other wireless applications:

- Transmission is one-way.
- The same signal is generally transmitted from neighbouring transmitter sites.

One-way transmission means that the peak-to-average power problem for OFDM is an issue only at the transmitter sites. Because there are relatively few of these, it has a very minor impact on system cost.

The second is somewhat more complex. With analogue TV transmissions, neighbouring transmitters must use different frequencies even if they broadcast the same signal. Otherwise, TVs that received more than one signal would exhibit a ghosting effect where multiple copies of a picture would appear on a TV screen. To overcome this problem, TV networks typically use a frequency repeat pattern of 11. Therefore, to transmit a single 8-MHz TV channel, 88 MHz of spectrum is needed so that neighbouring cells can be allocated different frequencies.

Digital transmission offers the possibility to overcome this and to reuse the same frequency on all sites, saving some 80 MHz of spectrum per channel. If the same frequency is used then the receiver will essentially see multipath signals; although, unlike in the mobile radio case, these will actually come from different transmitter sites. In principle this multipath could be combated using an equaliser. In practice, the high data rate transmission for digital TV (around 5 Msymbols/s), and the potential for path differences of up to 50 km between a nearby transmitter and a neighbouring one, result in path delays of something in the region of 800 symbols (compared to five for GSM). This is well beyond the complexity of current equaliser technology.

This leads to the use of OFDM. By using typically 2048 carriers, the symbol rate is reduced by 2048 and hence the multipath now extends only over up to one-quarter of a symbol. If each symbol is extended by this much then the multipath does not impact on the orthogonality of the signal, although the efficiency of the system is reduced slightly.

So, in summary, the receive-only nature of broadcasting and the desire to avoid using different frequencies on neighbouring sites makes OFDM well-suited to this application.

### W-LAN

W-LAN systems are characterised by very high data rates, in some cases approaching 20 Msymbols/s. They tend to be used in indoor environments where there are many multipath signals, but the time over which they are spread tends to be short as a result of the short distances involved. In such an environment, delay spreads are unlikely to extend much beyond six symbols. Nevertheless, the combination of high data rates coupled with mild multipath results in an equaliser requiring around 800 million multiplication operations per second. This is at the limit of current consumer technology and not appropriate for a W-LAN device intended to be low-cost.

As an alternative, OFDM with around 256 carriers has been proposed. The complexity of this approach is only around 10% of that of the SCM. Importantly, the resulting loss in performance is not critical. W-LANs generally make trade-offs towards lower cost because their shorter range allows them to be more wasteful with spectrum than cellular systems. Hence, the high data rates and the need for low-cost consumer equipment make OFDM suitable for high-speed W-LAN solutions.

*Fixed wireless*

There is a wide range of fixed wireless systems and so different solutions apply across the marketplace. In general, multipath within the fixed wireless environment is not severe. This is because directional antennas at the subscriber premises tend to reject multipath signals arriving at angles some distance off the main path. Simple geometry shows that these are the longer paths. The higher frequencies used also tend to require near line-of-sight transmission with the result that any reflected signals are at a much lower signal level than the main signal and hence cause less interference.

If we simplistically divide fixed wireless into consumer systems operating at 5 GHz or below and business systems operating at above 10 GHz, we can analyse the case for OFDM. For a consumer system with a 20-degree antenna beamwidth transmitting at around 3.5 Msymbols/s (corresponding to a 7 Mbits/s data rate for QPSK modulation) and with a range of up to 10 km, multipath occurs only over two symbols. The equaliser needed to combat this requires around 40 million multiply operations per second, well within the capabilities of DSPs. Hence, there is little reason to use the lower performing OFDM. Indeed, in the IEEE802.16a specifications on which this example is based, there are two suggested transmission modes, one based on SCM and the other on OFDM. Estimated performance figures given in that standard suggest that the SCM solution will slightly outperform the OFDM solution, delivering 7.5 Mbits/s compared to 7 Mbits/s in the same 6-MHz wide channel.

A business solution would have higher data rates, perhaps 20 Msymbols/s but with a narrower beamwidth of perhaps 10 degrees and a shorter range of around 2 km. For such a system the multipath delay is less than a symbol and hence there is no need for an equaliser. SCM systems here are both lower complexity and higher performance than OFDM solutions.

Another point is that fixed wireless systems tend to have relatively expensive RF components because they operate in frequency bands with limited economies of scale. Because OFDM places additional requirements on the RF amplifier it can add significantly to the cost of the subscriber equipment.

*Cellular*

Cellular systems tend to have relatively low data rates compared to the systems considered above but relatively long multipath delays as a result of the many reflections that occur in cellular environments. It is also important that they achieve the maximum possible performance as a result of the scarcity of radio spectrum. In 2G and 3G systems this has so far resulted in the selection of SCM, most recently based on CDMA, but some are now suggesting OFDM is more appropriate for 4G. Two proposals have been put forward which are worth considering separately. One, known as Flash-OFDM, is for a wide-area high-data-rate solution. The second, from the Japanese government, is the use of OFDM for a very high-data-rate '4G' solution.

Let us look at a generalised system similar to Flash-OFDM. Let us assume a symbol rate in the region of 1 Msymbol/s operating in a cellular band with a range of say 10 km. This is likely to generate multipath over around 20 symbols. An equaliser for this system would need around 100 million operations per second – viable but having a significant power consumption and cost. An OFDM approach would need only around four million operations per second. This is not a clear case – trading off the simplicity of OFDM versus the better performance of SCM is a difficult call. There is perhaps a space in the market for both solutions.

The Japanese proposal is for a system operating in the 3–6 GHz band and delivering 100 Mbits/s. In outline, any system in this band would have similar range and characteristics to the IEEE802.11a W-LAN solution which operates in the 5 GHz band. Indeed, the Japanese proposal is so close to 802.11a that it initially appears that they are re-inventing an existing solution. As already discussed, OFDM is appropriate for high-speed W-LANs so might well be appropriate for this system. But, importantly, this is more of a W-LAN solution than it is a cellular system.

**So is OFDM the New 'Technology of Choice'?**

The analysis above has demonstrated that:

- OFDM is appropriate for broadcasting because it enables the same frequency to be used on neighbouring transmitters.
- OFDM is appropriate for high-speed W-LANs because it reduces the complexity and therefore the cost.
- OFDM is inappropriate for fixed wireless because the use of directional antennas limits the impact of multipath.
- OFDM may be appropriate for cellular depending on the data rates and trade-off between cost and performance.

OFDM clearly has a role in certain situations. However, it is far from a universal access solution and its use needs to be determined based on the expected multipath coupled with the trade-offs being made in the system design. OFDM may be trendy but sometimes the boring old solutions are the best!

# References

[1] M Mouly and M-B Pautet, *The GSM System for Mobile Communications*, published by the authors, ISBN 2-9507190-0-7, 1992.
[2] L Harte, S Prokup and R Levine, *Cellular and PCS: the Big Picture*, New York, McGraw-Hill, 1997.
[3] W Webb, *The Complete Wireless Communications Professional*, Norwood, MA, Artech House, 1999.
[4] W Webb, *Introduction to Wireless Local Loop*, 2nd edn, Norwood, MA, Artech House, 2000.
[5] B Bing, *High Speed Wireless ATMs and LANs*, Norwood, MA, Artech House, 2000.
[6] G Held, *Data Over Wireless Networks: BlueTooth, WAP and Wireless LANs*, New York, McGraw-Hill, 2000.

# 5

# End User Demand

## 5.1 Why What the User Wants is Critical

Users will only subscribe to a service, or buy a new product, if they believe that they will benefit from it. They will only continue to use it, or pay for it, if their experience accords with their expectations. For a service to be successful it must both appeal to users before they buy it and provide them with benefits in excess of the costs after they have started to use it. A user will rarely be interested in the technical elegance of a solution. It is an often repeated truism that 'users don't care how a service is delivered'.

Understanding whether users will be interested in a product and whether they will buy it is difficult. Each user is different, and while segmentation can help to address this, it will only ever be an approximation. Users can be fickle, deciding on one day that they would like a new service, but then changing their mind later. Users rely on other users for reviews and recommendations, sometimes creating tipping points.

It is dangerous in predicting the future to rely entirely on user surveys. Users often do not appreciate the value of a new service until they try it – in the 1980s for example few felt that a mobile phone would be of value to them. Users often say one thing in a survey and then change their mind when they have to pay with their own money. Equally, it is dangerous to ignore the user – many new technologies have failed because they did not bring sufficient user benefits.

This chapter looks at what we can deduce from user behaviour and how this might affect our forecast of the future.

## 5.2 How People React to New Concepts

Even if the 'perfect' service is introduced, the entire population does not rush out to buy the service immediately. Instead, as is well chronicled, the service is first adopted by particular types of individual – the early adopters. Depending on their reaction it may then become

*Wireless Communications: The Future*   William Webb
© 2007 John Wiley & Sons, Ltd

adopted more widely. This is shown graphically in Figure 5.1, with some examples of various levels of device penetration in 2005.

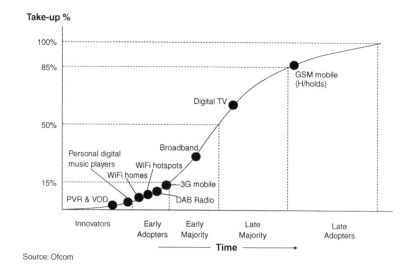

**Figure 5.1**  Technology adoption curve.

There are many complex dynamics here. The majority will typically adopt only if the experiences of the early adopters are good. They will rely on word of mouth, reviews and widescale promotion to convince them that the technology is worthwhile and mature. Initially pricing is often high as manufacturers seek to recover development costs and before economies of scale become important. Then it starts to fall as penetration increases. For technology products many now wait for falling prices as a signal to buy, having experienced situations in the past where they have bought products only to find the same or better products available for less shortly afterwards. There are two important implications:

- Much can go wrong. If the early adopters do not like the product, if the price is not set correctly, if the distribution chains do not champion it, if the consumers perceive a risk of early obsolescence, and so on, then the launch of the product will fail.
- The process will take some time. Regardless of how wonderful the product is, the majority will not adopt it until they feel enough time has passed for the product to become proven, for the price to stabilise and for it to be an acceptable thing for them to own. This is shown in Table 5.1.

Even the fastest diffusing product in this table – digital TV – has taken four years to reach mass market. Digital TV is a relatively simple proposition for the consumer. It is a one-off payment with no contract. The cost is low (around $100). The proposition is more TV channels – something readily understood. By comparison, broadband is much more complex. It requires a subscription, installation on a home computer and, without trying it first, the benefits of high speed may not be apparent.

**Table 5.1**  Time taken to reach mass penetration.

| Product | Launch date | Penetration | %/year |
|---------|-------------|-------------|--------|
| 2G | 1992 | 100% | ~ 10 |
| Digital TV | 2001 | 60% | 15 |
| Broadband | 2000 | 30% | 6 |
| 3G | 2002 | 15% | 5 |
| DAB radio | 2000 | 10% | 2 |
| WiFi homes | 1999 | 10% | 2 |

From this we can conclude that a new service or product might take anything between four and ten years to reach mass adoption, even if the service is perfect in every respect and there is a strong user demand for it. Services for which the benefits are less familiar to the end user will take longer. Add five years for the standardisation and development of a new technology and it might take 15 years from conception to large-scale success. This is the truth of 'the fast moving world of telecommunications' that is often not understood widely.

## 5.3  Changing Patterns of Spending

Those launching new services typically assume that people will pay more for them. There are some cases where the services are substitutes and those offering them are trying to persuade users to change their spending from one service to another but most advances assume additional spending. However, consumers do not have infinite resources, and even if they strongly desired a particular service they will not adopt it if they cannot afford it.

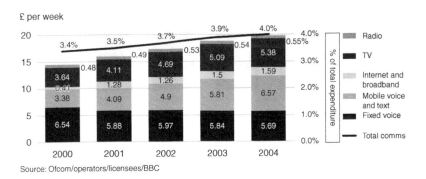

**Figure 5.2**  UK consumer spending on telecommunications.

Figure 5.2 shows how UK spending on communications has changed over the last five years, both as an absolute spend (left-hand axis) and as a percentage of total household spending (right-hand axis). There are some interesting trends. The decline in traditional fixed line voice is clear as people switch to mobile phones. Internet and broadband subscriptions have grown around four-fold over the period and subscription TV has experienced strong growth. Overall, the change in household spending has been 0.6% over four years, or around 0.15% a year, reaching £20/week by 2004 (around $36). This equates to weekly spending levels growing by around 75p or $1.35 a year.

There are many reasons why the change in spending has shown relatively smooth growth. Some are to do with the rate of adoption, but others are to do with household budgeting. Average household income rises only slowly and typically only slightly above inflation. Hence, disposable income, on average, changes very slowly. If users are to increase spending on communications they must typically decrease spending in some other area. This may take some time as they cancel subscriptions, make investments that will reduce overall spending or just become comfortable with a changed lifestyle. As well as there being slow growth, there will be a maximum level of spend. Users need to buy food, clothes, fuel, holidays and many other things. Just what percentage of spend communications can rise to is unclear, but it will probably be less than 10%.

So, in summary, it is unlikely that total communications spending will grow by more than 0.15% of household income per year – equating to weekly spending changing by $1.35 per year. Further, this level of growth cannot be indefinite. Many different communications needs compete for this growth – including mobile, broadband, TV and radio – and households will divide the growth among some or all of these. So for a particular service, such as mobile, growth may be less than the 0.15% level, unless substituting for another service.

Turning this into something more understandable, the average revenue per user (ARPU) per month for the UK is shown in Figure 5.3. It has been reasonably stable at around £17 ($30.5) per month for some time.

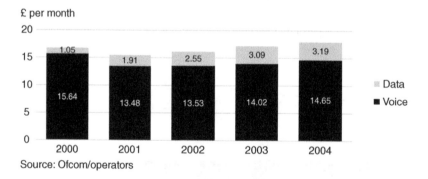

**Figure 5.3**   ARPU.

If, say, 50% of the available increase in spending was allocated to mobile, then this would be around 37p/week or about £1.55/month ($0.66 and $2.80, respectively). This would allow a 9% annual growth in ARPUs. Of course, it is possible that substitution might increase this further, or that more than 50% of the available increase could be allocated, so this is not a hard limit. Nevertheless, a service requiring annual ARPU growth in excess of 9% appears unlikely, and even 5% would be challenging.

So not only do users take some time to adopt new services, even 'perfect' ones, they may not be able to afford them, or may take time to modify spending patterns to free up money.

## 5.4 What they have Today

The typical office worker in a developed country currently has a wide range of different communications types, including:

- the office telephone, used mostly for voice communications complete with voice mailbox
- the office fax machine, now being less used as email takes over
- the office local area network (LAN), providing high data rate communications such as email and file transfer, perhaps wirelessly
- a range of services for working when out of the office, including dial-up networking, high-speed networks in hotels and increasingly WiFi hotspots
- mobile telephones providing voice communications, a mailbox and in some cases email and low-speed data access
- a home telephone providing voice communications and dial-up access along with a home answering machine
- a computer at home linked to a different email system, perhaps using high-speed connections such as Asymmetric Digital Subscriber Line (ADSL) or cable modems
- a home wireless network using WiFi enabling computers and other devices to be connected together
- broadcast TV and radio services provided across a range of bearers including terrestrial analogue and digital broadcast, cable, satellite and via the Internet.

Managing all these different communication devices is complex and time-consuming. The worker who has all of these will have four 'phone' numbers, three voice mailbox systems and two email addresses. There is typically little interconnection between any of these devices so that all the different mailboxes have to be checked separately, using different protocols and passwords. At best, one mailbox might be forwarded to another. Contacting such an individual is problematic due to the choice of numbers to call, the simplest option often being to call the mobile number as the one most likely to be answered. Although many are working on systems such as unified messaging, designed to allow all types of communications (voice, fax, email) to be sent to one number, we are still some way from the ideal situation where individuals only have one 'address' and all communications is unified. Effectively, there remains little convergence, at least as far as the user is concerned, between all these different methods of communications.

People also utilise wireless communications in the home and office for applications such as remote-control toys, garage door openers, contactless smart-card entry and payment systems and even TV remote controls which use part of the electromagnetic spectrum, albeit at rather higher frequencies than radio. Most of these devices are uni-functional. That is to say that the garage door opener performs only one function, the opening of the garage door. Consolidated remote controls able to control a range of devices have proved popular and it may be that over time devices such as PDAs will learn to control multiple devices around the home.

## 5.5 What they want Now

In 2006, for the travelling businessman, one of the key issues to contend with during the day and in the evening was how to check emails. Using the mobile phone was generally certain (given adequate coverage) but slow and often expensive. Finding a telephone socket in the hotel sometimes involved scrabbling around on the floor, followed by a skim through the telephone guide to try to understand whether a '9', a '0' or some other prefix was needed to get an outside line. For those with newer laptops, finding a W-LAN hotspot was another possibility, but one that sometimes required time spent attempting to configure the laptop, although with the integration of W-LAN functionality into Windows XP this was improving rapidly. Automatic retrieval of emails by the fastest and cheapest method still seemed a distant dream.

From this description we might conjecture that a businessman and perhaps others would value unification of all their different voicemails and email accounts, with a single address. However, this would lead to the potential for work and home life to become confused, and so some form of intelligent filtering and call redirection based on preferences, location etc. would be needed. A user would also value the ability to simply connect to the most appropriate network wherever he or she was with little action needed. It would be nice to have information based on location and context – such as travel information delivered automatically. We also think users might value video communications, which we discuss further in the box.

---

### Do people really want video?

Video communications has often been talked about but consistently failed to become a commercial reality. Is this because people do not actually want video communications or because the technology is insufficiently mature? This is an important question for the future of wireless communications since without video communications one of the key drivers for advanced wireless networks is removed.

*Commercial and technological issues*
There have been commercial problems with previous videophone solutions, all of which have been based around fixed line communications. Early dedicated video systems generally worked at a bandwidth of about 9.6 kbits/s with a resulting small and jerky picture. Equipment was expensive and dedicated solely to video telephony. As with all systems requiring dedicated equipment, there was a start-up problem in that there was little point being one of the first people to buy a video phone since there was nobody else to call (see also Metcalfe's law in Chapter 6). This is a huge problem for any new technology requiring dedicated receiving equipment and probably sufficient to kill this particular videophone concept even had the picture quality been acceptable.

Video telephone has also become possible over the Internet. This has some benefits over the earlier dedicated systems in that mostly standard equipment could be used – all that was required in addition to a computer and modem was a relatively cheap webcam. However, in 2006 the Internet was still poor at delivering packets consistently and with minimal delay, with the result that the quality was generally inadequate. Nevertheless, Skype and others were planning significant video initiatives which might succeed in overcoming some of these problems.

For a video call technology to be acceptable it ideally needs to:

- use equipment that is already widely available such that there are many other potential people to call
- be of a sufficient quality that it is not perceived as jerky or blurred by the users
- be sufficiently inexpensive that the addition of video to a voice call is perceived as worthwhile;
- be simple to use, ideally no more complex than setting up a voice call.

***Do people actually want it?***
Many, today, comment that video communications is pointless. They note the difficulties in positioning the camera, the fact that they may not want others to see them and that voice communications has proved perfectly acceptable for decades. Perhaps, they concede, it might be nice for the grandparents to see their grandchildren, but otherwise there is limited need. It is rare to find someone who enthuses over how nice it would be to always be able to see the person at the other end of the call. On the basis of a simple poll the desire for video communications would seem low.

But perhaps there is an issue of familiarity at play here. People have become used to voice communications and learnt how to change their behaviour in order to communicate effectively without visual cues. For someone familiar with voice communications, as most are, it seems perfectly acceptable. But then, so was black and white television before colour emerged, and the video recorder before DVD players became inexpensive. Just as we perceive the world in colour and so naturally prefer colour television, so we communicate both visually and verbally and so should prefer video communications in most circumstances.

Without hard evidence, any premise that people actually want video communications must remain an act of speculation. Here we are of the view that if the technological challenges can be overcome and the networking problems avoided by seeding the population with video-capable devices, then many might be surprised at how readily they adopt the new technology and how they dislike it when they have to revert to voice-only communications. But we accept that this is an act of faith on our behalf!

Particularly for the non-commercial user, the idea of being able to 'post' their own information to share with others appears to be becoming increasingly important. This is encapsulated in the idea of Web 2.0, and visible in the success of sites such as MySpace. This may result in increases in data traffic, particularly in the uplink, which is often thought to be less utilised by many applications. This will depend on the extent to which users wish to upload information while on a mobile, rather than fixed, connection.

## 5.6 Security, Privacy and Health Concerns

The use of communications systems, whether wireless or wired, can lead to many unwanted effects. Users can have their transmissions eavesdropped, or sensitive information tapped.

Spam and other unwanted messages can make communications more difficult for the user. Viruses can reduce the functionality of devices. Additional wireless information, such as location, can be seen as an intrusion on privacy. Specifically to the mobile phone, there can be concerns about its impact on health.

Most of these problems have already occurred on fixed communications networks. Mechanisms are available for dealing with viruses and spam and there are security solutions that prevent eavesdropping. Applying similar approaches to wireless communications would seem mostly straightforward, although it may take users some time to realise that they need a virus checker for their phone.

The use of location data raises different concerns. For example, in 2005 Benetton backed down on plans to fit RFID tags to store merchandise because of customers' concerns that they could then be tracked. Cellular operators have been storing a coarse form of location information for some time and users have not broadly been concerned by this. It seems likely that most users will accept the ability to monitor their location in more detail as long as they perceive it brings them benefits while the information itself is made available only to responsible parties. Manufacturers are also working on mechanisms to provide users with more control – for example RFID tags where the corner can be snapped off by the user, dramatically reducing the range of the tag to a few centimetres, thus preventing tracking but still allowing a store to read details at the counter.

Health concerns have surrounded mobile phone usage since the late 1980s. Many detailed studies have been made, which have mostly shown that there appears to be little evidence that there are any health effects, but have generally been unable to prove conclusively that there will never be any. Health concerns erupt periodically, although the frequency of these does seem to be diminishing. As the length of time that people have been using mobiles increases, with little apparent evidence of any increase in related diseases, it seems likely that these concerns will slowly diminish.

## 5.7 The Handset Subsidy Problem

In many countries cellular handsets are subsidised by cellular operators. Users are able to obtain handsets free, or at little cost, in return for some form of contract. This has an impact on the sale of more complex or converged handsets. If the cellular operators decide not to offer, for example, handsets with integrated GPS receivers, then these will be available only unsubsidised, which will make them appear unreasonably expensive to many consumers. Further, subsidised handsets are mostly sold through retailers owned by cellular operators. Typically the focus of such retailers is phones, rather than for example GPS receivers. Hence, they will tend not to stock items that are GPS units with accompanying phone, rather than the other way around. Their understanding of the GPS functionality may also be poor compared to the phone functionality – in any case the phone itself will likely be complex and their ability to demonstrate all the functionality may be weak. Finally, there may be functions that the cellular operator perceives might reduce revenue, such as the ability to make voice calls over W-LANs. Operators may choose not to stock such phones or to decrease the subsidy that they provide.

The other option open to consumers is to buy a stand-alone GPS unit. This will be sold in electronics stores where they have the skills to demonstrate it. The functionality will also be relatively easy for the consumer to understand.

Hence, in various ways, the current approach to selling handsets will tend to militate against the take-up of complex, multi-functional or converged phones. As the phone market gradually becomes more mature, the degree of subsidy may slowly decrease but it will be some considerable time before it is removed sufficiently to reduce its impact on handset sales.

## 5.8 In Summary

In this chapter we have stressed several points.

- For a new service to be accepted it must be something that the users actually want, but users often cannot envisage future possibilities, so simply asking users might not deliver the right answer.
- Users take time to adopt new services and ideas. It can take between four and ten years for even the most 'perfect' service to become widely adopted, and there are many pitfalls on route that can extend this, or even cause the service to fail.
- Users have limited resources to spend on communications and take time to change their spending habits. Any expectation of revenue growth in excess of around $2.80/month happening in less than a year is likely to be unrealistic, unless there are unusual circumstances or conditions.
- The industry structure, including handset subsidy, tends to work against the introduction of advanced or novel handsets.
- Users today have a wide range of communications offerings, but this range itself causes problems of multiple numbers, mailboxes etc. Simplification of this would bring benefits.
- Services that would bring benefits to users in general are those that make communications ubiquitous, simple, inexpensive and unrestricted in format, data rate etc.

The next chapter takes a look at technological evolution and asks whether it is likely to bring about the service improvements listed here at a reasonable cost.

# 6

# Technology Progress

## 6.1 Technology is a Critical Input to any Forecast

Many of the advances in wireless communications to date have been enabled by technological progress. Improvements in processing power have enabled more complex radio technologies to be deployed and radio equipment to become much less expensive. The change from the early 1980s, when few had cellphones, to today where they are ubiquitous has been made possible almost entirely by advances in technology, providing greater capacity and lower cost. The direction taken by technology in the future has the potential to significantly impact upon the success and type of future communications systems. This chapter discusses how technology will evolve over the next 20 years and the implications for forecasting the future of wireless. The different areas of technology that interest us are:

- Technologies enhancing the efficiency of transmission. These allow more people to communicate, using higher data rates, at lower costs.
- Technologies lowering the cost of networks or handsets, making available to the mass market services that were previously too expensive.
- Technologies enhancing the way we interact with terminals, including screens, keyboards and speech recognition.
- Technologies leading to 'artificial intelligence' enabling devices to predict our behaviour and hence provide context-aware information.
- Compression technologies, enabling information to be sent in less bandwidth and stored in less memory.

Some technical fundamentals, such as enhancements in processing powers, underlie many of these. Others, such as display technology, are specific to particular categories. In this chapter we first look at the more fundamental underlying issues. We have divided these into two categories: 'true' laws (those derived mathematically or scientifically),

*Wireless Communications: The Future*   William Webb
© 2007 John Wiley & Sons, Ltd

and 'empirical' laws (those based on observation of trends). We then take a look at technological developments that can be foreseen in each of the areas listed above before drawing conclusions.

## 6.2 Key Technical Fundamentals: The 'True' Laws

Of all the true laws covering wireless communications, Shannon's law stands out as being most relevant [1, 2]. This law states that, under certain assumptions, the maximum information measured in bits that can be transmitted per second per hertz of radio spectrum, $C$, is:

$$C = \log_2(1 + SNR) \tag{6.1}$$

where SNR is the signal to noise ratio at the receiver. Practical systems running over wired connections can now come very close to the Shannon limit. However, Shannon alone cannot tell us what the maximum capacity of a wireless network will be given a fixed amount of spectrum. This is because multiple radios can use the same piece of spectrum, generating interference to each other. If there are few of them then each radio has a high SNR and hence can transmit a relatively large amount of information, but with few radios then the total information transmitted across the network is small. If there are many of them, then the interference is high, so the SNR, or more correctly the 'signal to interference plus noise ratio' (SINR), is low and the information per radio is less. But there are more radios and hence the total information sent across the network may be higher. The key to applying Shannon to discover the maximum capacity of a piece of spectrum is to find the point where the combination of information rate and number of users is at a maximum.

This can be done by making some basic assumptions about the geometry of a cellular system and then performing simulations to determine where the maximum lies [3, 4]. It turns out that the maximum occurs towards the point where there are more users, but more interference – in fact at an SINR of around 6 dB. Using this information and manipulating the Shannon equation, the number of channels, $m$, that can be supported within a piece of spectrum, can be derived as:

$$m = 1.42\alpha \frac{B_t}{R_b}. \tag{6.2}$$

where $B_t$ is the total available bandwidth, $R_b$ is the user bit rate and $\alpha$ is a factor relating to the closeness to which the Shannon limit can be approached.

If we could build a perfect radio system which met the Shannon limit, then $\alpha$ becomes 1. Hence, if we have 1 MHz of spectrum and each user's 'call' requires 10 kbits/s of data, then we can accommodate approximately a maximum of 142 voice calls per cell in a clustered environment where spectrum is reused from cell to cell. Third-generation systems realise around 60–100 'calls' per carrier per sector. Each carrier is 5 MHz wide, so the number of 'calls' per sector per megahertz is around 12–20. By using enhancements such as HSDPA, levels of around 30–50 calls per cell can be attained.

So current systems deliver about a third of the maximum capacity achievable. This is not unreasonable. Shannon assumes that any interference takes the form of white noise – the most benign type for most systems – whereas mobile radio channels are considerably more hostile, with fading and multipath propagation [5]. Overcoming these problems requires equalisation, channel estimation and error-correction coding which typically reduces the efficiency to around half of its original levels. Other problems such as device imperfections, complexity of filtering, inability to implement perfect receivers, and so on, all add to the difficulty in reaching this limit.

An important point of note about this limit is that it is stated in terms of capacity per cell. Simply adding more cells, perhaps by replacing one macrocell with a handful of microcells, will dramatically increase the capacity. Given the relative ease of doing this, and the apparent difficultly in making dramatic capacity gains through technology, it seems likely that smaller cells and not better technology will be the key path to increased radio capacity in the future.

*Does MIMO invalidate the Shannon limit?* MIMO systems work by having a number of antennas at the base station and a number at the subscriber unit. A different signal is transmitted from each antenna at, say, the base station, but all transmissions are at the same time and same frequency. Each antenna at the subscriber unit will receive a signal that is the combination of all the transmissions from the base station modified by the parameters of the radio channel through which each passes. In a diverse environment, each radio path might be subtly different.

If the characteristics of each radio channel from each transmit antenna to each receive antenna are known then, mathematically, it is relatively simply to apply a 'matrix inversion' operation and deduce what data was transmitted from each antenna. Each path from one antenna to another conforms to the Shannon limit, but the combination of all the paths can rise many times above it.

In order for MIMO to work, there must be a reasonable degree of difference between each of the transmission paths – if they are all identical then each of the signals received at the subscriber unit antennas will also be identical and it will not be possible to invert the matrix and hence extract the different transmitted signals. Differences across paths tend to occur in mobile environments where there is rarely a line of sight to the base station and multiple reflections take place. However, it is also necessary that each of the transmission paths is known with high accuracy – otherwise the solution to the matrix equation will be inaccurate. In a typical mobile environment, the transmission channels are constantly changing their parameters as the mobile moves, or vehicles or people move in the vicinity. Finding out what these parameters are requires 'sounding' the channel – sending a known sequence from the transmitter and using the knowledge of this sequence at the receiver to deduce what the parameters of the channel must be. Sending this sequence is an overhead that reduces the capacity of the system to send the user information. If the channel changes rapidly, then it must be sounded frequently, reaching a point where any gains from multiple antennas are negated through lost time sounding the channel.

In the fixed environment, tracking the channel parameters is possible but there is a higher likelihood of a line of sight, which reduces the channel diversity to a point where MIMO may not work.

Added to all of this is that additional cost of MIMO systems and the need for multiple antennas which are unsightly at base stations and difficult to integrate into handhelds.

We draw the tentative conclusions that MIMO will not result in significant capacity gains over those predicted by Shannon. Further issues surrounding MIMO systems are discussed below.

## 6.3 Key Technical Observations: The 'Empirical' Laws

Almost all engineers are aware of 'Moore's law' – the phenomenon it describes has had an amazing impact on electronics over the last few decades, continually making devices cheaper, better and smaller. Moore's law is not founded in laws of nature, like Ohm's law for example, but is essentially an economic prediction based on observations of previous trends in scientific research and industry.

In this section we take a look at a range of empirical laws, in particular those with relevance to communications technology, understanding where they have come from and what their impact might be in the future.

### 6.3.1 Moore's Law

We start with by far the most famous – Moore's law. In 1965, Moore predicted that the number of transistors the industry would be able to place on a computer chip would double every year. In 1975, he updated his prediction to once every 18 months. Most recently in 2003, he suggested that at least another decade of growth was possible (see Figure 6.1).

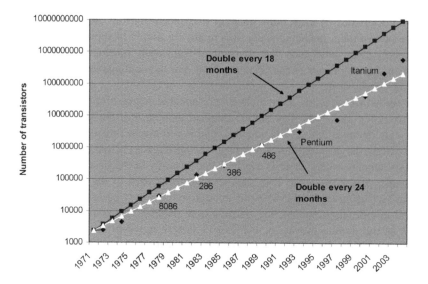

**Figure 6.1**   Growth in Intel processors since 1971.

Moore was born in San Francisco, California, on 3 January 1929. He earned a BS in Chemistry from the University of California at Berkeley and a PhD in Chemistry and Physics from the California Institute of Technology in 1954.

He joined Caltech alumnus William Shockley at the Shockley Semiconductor Laboratory division of Beckman Instruments, but left with the 'Traitorous Eight' to create the influential Fairchild Semiconductor Corporation.

He co-founded Intel Corporation in July of 1968, serving as Executive Vice President until 1975 when he became President and Chief Executive Officer. In April 1979, Dr Moore became Chairman of the Board and Chief Executive Officer. He currently serves as Chairman Emeritus.

He is a director of Gilead Sciences Inc., a member of the National Academy of Engineering, and a Fellow of the IEEE. Moore also serves on the Board of Trustees of the California Institute of Technology. He received the National Medal of Technology from President George Bush in 1990. The library at the Centre for Mathematical Sciences at the University of Cambridge is named after him and his wife Betty.

The end or 'death' of Moore's law has often been predicted. Several times in the past, it appeared that technological barriers such as power consumption would slow or even halt the growth trends, but engineers have always found ways around the barriers. Now, however, a fundamental barrier is emerging —the technology is approaching atomic dimensions, raising all sorts of new challenges. Best industry predictions at present suggest that gains will slow around 2010 and may stop altogether around 2016 when atomic levels will be reached. Use of multiple parallel processors may allow some further improvements, but parallel systems are costly, power hungry and difficult to work with.

There have been some suggestions that the close way that industry has delivered on Moore's law over the years is something of a self-fulfilling prophecy. It has been argued that because engineers expect their competitors to achieve Moore's law growth they continue to innovate until they achieve it themselves. Whether there is any truth in this or not, Moore's law has been one of the most dramatic trends in modern society.

## 6.3.2 Metcalfe's Law

Metcalfe's law states that the value of a network equals approximately the square of the number of users of the system ($n^2$).

Robert Metcalfe was born 1946 in New York. He is credited with inventing the Ethernet. He also co-founded 3Com.

He graduated from MIT with two Bachelor's degrees, one in electrical engineering and another in industrial management. He earned his PhD from Harvard with a thesis on packet switching.

*(continued)*

Metcalfe was working at Xerox in 1973 when he invented the Ethernet. In 1979 he left PARC and founded 3Com, a manufacturer of networking equipment. In 1990 he retired from 3Com and became a publisher and journalist, writing an Internet column. He became a venture capitalist in 2001 and is now a General Partner at Polaris Venture Partners.

Metcalfe received the National Medal of Technology in 2005, 'for leadership in the invention, standardisation, and commercialisation of Ethernet'. In 2004 there were around 200 million new Ethernet ports shipped worldwide.

The law is often illustrated with the example of fax machines or video telephones. A single fax machine is useless, but the value of every fax machine increases with the total number of fax machines in the network, because the total number of people with whom you may send and receive documents increases.

More recently, some researchers have suggested that Metcalfe's law might significantly overestimate the value of adding connections. They suggest that the value of a network with $n$ members is not $n$ squared, but rather $n$ times the logarithm of $n$. Their primary justification for this is the idea that not all potential connections in a network are equally valuable. For example, most people call their families a great deal more often than they call strangers in other countries, and so do not derive the full value $n$ from the phone service.

The impact of Metcalfe's law may be more important for regulators and innovators than others. It shows that initially a new network has very little value but as it grows the value grows faster than the number of nodes. Therefore, the benefits to society of another person joining the network may be greater than the benefits to that individual. Hence there may be situations where an individual decides not to join the network because the value to him/her as an individual is too low, but the value to society would make it worthwhile. This is sometimes called an 'externality' by economists who would recommend that intervention should take place, perhaps in the form of subsidisation, to increase the number of people on the network.

Unlike Moore's law, Metcalfe's does not have a time limit to it. It will likely apply to a wide range of new networks in the future as new types of devices and networks are invented.

## 6.3.3 Gilder's Law

George Gilder noted that network bandwidth (in fibre cables and similar) tripled every 9–12 months and predicted around the year 2000 that this would continue for at least 25 years. The impact of this, he predicted, would be that Internet traffic would double every 100 days.

George Gilder was born in 1939 in New York City. He studied at Harvard University under Henry Kissinger and helped found Advance, a journal of political thought. In the 1960s Gilder served as a speechwriter for several prominent officials and candidates, including Nelson Rockefeller, George Romney and Richard Nixon. In the 1970s, as an independent researcher and writer, Gilder began an excursion into the causes of poverty, which resulted in his books *Men and Marriage* (1972) and *Visible Man* (1978); and hence, of wealth, which led to his best-selling *Wealth and Poverty* (1981).

Gilder pioneered the formulation of supply-side economics when he served as Chairman of the Lehrman Institute's Economic Roundtable, as Program Director for the Manhattan Institute, and as a frequent contributor to A. B. Laffer's economic reports and the editorial page of *The Wall Street Journal*.

The investigation into wealth creation led Gilder into deeper examination of the lives of present-day entrepreneurs, culminating in many articles and a book, *The Spirit of Enterprise* (1986). That many of the most interesting current entrepreneurs were to be found in high-technology fields also led Gilder, over several years, to examine this subject in depth. In his best-selling work, *Microcosm* (1989), he explored the quantum roots of the new electronic technologies. A subsequent book, *Life After Television*, was a prophecy of the future of computers and telecommunications and a prelude to his book on the future of telecommunications, *Telecosm* (2000).

Gilder's law, to some degree, prefaced the problems suffered by telecoms companies, particularly those laying fibre, in the late 1990s and early 2000. Network capacity grew extremely rapidly as dense wavelength division multiplexing (DWDM) techniques were applied to fibre. Capacity grew faster than the rise in traffic levels, leading to a glut of transmission capacity, falling prices, and financial problems for a range of companies such as Global Crossing.

To some degree, market forces have slowed Guilder's law in that there has been little incentive to devise and invest in techniques for enhancing fibre capacity when there is already excessive capacity. Instead, engineers have come to think of the capacity of networks as limitless, assuming that new techniques to enhance fibre capacity will become available as they are needed.

## 6.3.4 Cooper's Law

Marty Cooper noted that, on average, the number of voice calls carried over radio spectrum has doubled every 30 months for the past 105 years. The effectiveness of spectrum utilisation in personal communications has improved by a factor of about a trillion since 1901. He predicted this would continue for the foreseeable future.

A pioneer in the wireless communications industry, Cooper conceived the first portable cellular phone in 1973 and led the ten-year process of bringing it to market.

During 29 years with Motorola, Cooper built and managed both its paging and cellular businesses and served as Corporate Director of Research and Development.

*(continued)*

Upon leaving Motorola, Cooper co-founded Cellular Business Systems Inc. and led it to dominate the cellular billing industry with a 75% market share before selling it to Cincinnati Bell.

Cooper then founded ArrayComm in 1992, growing it from a seed-funded start-up into a world leader in smart antenna technology.

Cooper received the American Computer Museum's George R. Stibitz Computer and Communications Pioneer Award in 2002; he was an inaugural member of RCR's Wireless Hall of Fame; *Red Herring* magazine named him one of the Top 10 Entrepreneurs of 2000; and Wireless Systems Design gave him the 2002 Industry Leader Award.

The gains predicted by Cooper come from a range of sources, but most significant is the reuse of frequencies in smaller and smaller cells. When Marconi conducted the first radio transmissions in 1895, the energy from his spark gap transmitter occupied the entire usable radio spectrum. The first transatlantic transmission in 1901, which blanketed an area of millions of square miles, was capable of sending only a few bits per second. In fact, only a single such transmission could be accommodated on the surface of the earth using that technology.

Focusing on the most recent period, Cooper suggests that the number of voice calls has improved a million times since 1950. Of that million-times improvement, roughly 15 times was the result of being able to use more spectrum (3 GHz vs 150 MHz) and five times was from using frequency division, that is, the ability to divide the radio spectrum into narrower slices (25-kHz channels vs 120-kHz channels). Modulation techniques (such as FM, single sideband, time-division multiplexing, and various approaches to spread spectrum) can take credit for another five times or so. But the lion's share of the improvement – a factor of about 2700 – Cooper suggested was the result of effectively confining individual conversations to smaller and smaller areas by spatial division or spectrum reuse.

Even tighter geographical reuse, to a distance of only a few feet, would increase the effectiveness of spectrum use by 10 million times over today's capabilities. If we approached that 10 million times improvement at the current rate of doubling every 2.5 years, it would take 58 years, at which time we would have the capability of delivering the entire radio frequency spectrum to every single individual.

Cooper's law tells us that despite being close to the Shannon limit for a single channel, there is no end in sight for practical increases in wireless transmission. Ever higher data rates and capacities are entirely possible if we are prepared to invest in an appropriately dense infrastructure. They also tell a regulator that providing additional spectrum is not the only way to increase capacity, although it does not provide any guidance as to which would be the most economic

Cooper's law is very difficult to verify. While the start point in 1901 is not in much doubt, any other evidence is hard to come by. Figure 6.2 is based on a few limited pieces of data:

- *An estimate of the total cellphone voice call minutes in 2001 worldwide.* This gives us a fixed point on the chart. However, it does ignore wireless capacity provided via cordless phones, W-LANs, and other systems.

- *The growth in cellphone subscriptions between 1981 and 2005.* As a first approximation we can assume that voice calls per user stayed approximately constant on average over this period.

If we use subscription levels to extrapolate from our known point in 2001, and then in a very cavalier fashion assume constant growth from 1901 to 1981, we get the chart in Figure 6.2.This suggests that Cooper's Law has overstated growth between 1900 and 1980 but understated it subsequently.

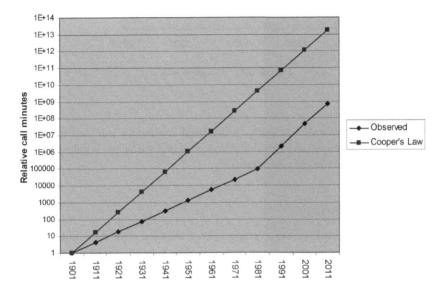

**Figure 6.2**   Cooper's law compared to observed trends.

## 6.3.5 Edholm's Law

Edholm sets out three categories of communications – wired or fixed, wireless and nomadic. Nomadic is a form of wireless where the communicator is stationary during the period of communications. According to Edholm's law, data rates for these three telecommunications categories increase on similar exponential curves, the slower rates trailing the faster ones by a predictable time lag.

Figure 6.3 shows data rates plotted logarithmically against time. When drawn like this, it is possible to fit straight lines to each of the categories. The lines are almost parallel, although nomadic and fixed technologies gradually converge at around 2030. For example, in 2000, 2G delivered around 10 kbits/s, W-LANs connected to dial-up delivered 56 kbits/s, and the typical office local area network (LAN) provided 10 Mbits/s. Today, 3G delivers 300 kbits/s, a home wireless LAN is about 50 Mb/s and typical office LAN data rates are 1 Gbits/s. Edholm's law predicts that, in 2010, 3G wireless will deliver over 1 Mbits/s, Wi-Fi around 200 Mbits/s and office networks about 5 Gbit/s.

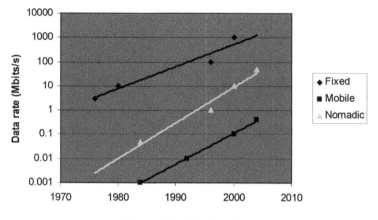

**Figure 6.3**   Edholm's law.

Edholm's law overlaps with Guilder's on the fixed bandwidth side and to some degree with Cooper's on the wireless side. Key is its prediction that wired and wireless will maintain a near-constant differential in data rate terms.

### 6.3.6  Growth in Disk Size

Figure 6.4 shows the growth in the average size of the hard disk in a mid-range PC. Of course, much larger disks are available in high-end PCs or as additional external drives.

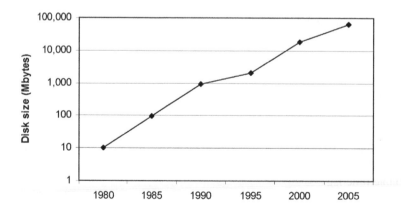

**Figure 6.4**   Average hard disk size on a PC.

What is clear from this chart is that when disk size is plotted on a logarithmic scale against time there is an almost straight-line relationship, although the growth did slow a little in the mid 1990s before picking up again near 2000. Simplistically, we might assume that this trend continues onwards throughout the period of interest, and using the trend line from this chart as the basis we can extrapolate to gain the graph shown in Figure 6.5.

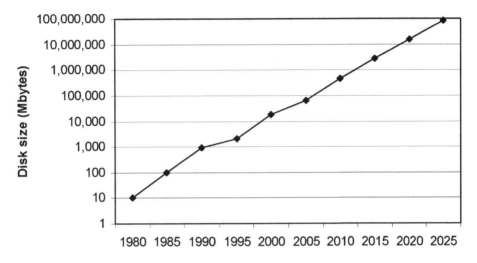

**Figure 6.5**  Prediction of future hard disk sizes.

On this basis we might expect hard disk size to rise by around an order of magnitude every six years, getting to around 15 Tbytes (15 000 Gbytes) by 2020. As with Moore's law some have suggested that we are reaching the end of the growth curve as the magnetic domains on the disk become too small to reliably hold their magnetic alignment, but others suggest that vertically stacked magnetic elements or different materials might be expected to solve this problem. As an interim solution, it is always possible to simply build a stack of lower capacity disks.

Hard disk sizes increase only because users have a need for the capacity, although there is something of a circular issue here since as hard disk sizes increase, software producers write programs able to take advantages of the increased size, with larger files which in turn spur further growth of hard disk space. By 2003, video editing capabilities were increasingly becoming standard on PCs. An hour of captured video requires of the order of 20 Gbytes of storage (although it is then typically compressed for writing to a DVD or similar). It would not be unreasonable to have tens or even hundreds of hours of captured video. At first, these will be archived to other storage media, but as time goes on, they will increasingly reside on computers.

While interesting, it is not clear what the result of this trend will be. Some, including the predictions we made in 2000, suggest that as hard disks grow, so will file sizes and that users will wish to exchange these. Hence, data rate requirements on transmission networks might grow at around the same pace as disk capacity. Others have predicted almost the opposite. They suggest that as hard disks become larger, computers, iPods and similar devices will come pre-loaded with the complete set of all songs, books or films, ever made. Users will only need to download material produced since the date of manufacture. While this has some logic, much of the material currently downloaded is newly produced, or produced by the user themselves. Hence, in practice, pre-loading may have little effect on file transfer requirements.

### 6.3.7 Goodhart's Law

Goodhart's law has nothing directly to do with technology. The original form of Goodhart's law states 'As soon as the government attempts to regulate any particular set of financial assets, these become unreliable as indicators of economic trends.' Broadly, this occurs because investors determine the likely impact of the government action and invest in a manner to benefit from it. This changed investment behaviour changes the key relationships.

The Bank of England had its own version of the law, which was 'Any observed statistical regularity will tend to collapse once pressure is placed upon it for control purposes.'

Professor Strathern further re-stated Goodhart's law more succinctly and more generally as 'When a measure becomes a target, it ceases to be a good measure.'

Apart from the somewhat amusing inference that no economic measures are likely to be particularly valuable if probed too deeply, Goodhart's law does suggest that any of these empirical laws, if widely believed, are likely to alter behaviour.

### 6.3.8 Laws or Trends?

Forecasting the future is essential in business and academia. All decisions to invest in research, product development or network deployment are based on an expectation of growth or change. Trends are clearly important predictors of the future – but there can be circumstances that cause them to break down and in many cases growth cannot continue for ever. 'Laws' such as those we have mentioned are a form of trend where the future continuation has been closely examined by an industry guru and appears to them to have a strong likelihood of continuation.

But as we saw with Goodhart's law, things are not quite as simple as that. If enough people believe a particular law – perhaps because of the eminence of its originator – they may change their behaviour as a result. This might either reinforce the law (potentially as we saw with Moore's) or cause it to break down (as we saw with Guilder's). Laws of this kind, then, need to be treated with considerable caution.

## 6.4 Technologies on the 'Radar Screen'

Technologies typically take many years to come to fruition. It was around 17 years after the first exploratory talks on 3G that networks were commercially deployed. OFDM was first patented in the 1960s but implemented in the late 1990s. Smart antennas have been a research topic for over a decade and broadly remain so. Ultra wideband was discussed many decades ago. Research started in earnest around 1995 and yet by 2005 there were still no commercial UWB products available. Hence, an understanding of products currently in research or development can provide a good idea of what technologies might be available over the next 20 years. We have segmented this section in line with the different technology areas identified above. In some areas we have called upon subject matter experts to provide contributions.

## 6.4.1 Technologies Enhancing the Efficiency of Transmission

### Software-defined Radio

Many future visions of wireless communications involve multi-modal devices connecting to a wide range of different networks such as 2G, 3G, WiFi and BlueTooth. They may even involve devices modifying their behaviour as they discover new types of network or as home networks add additional functionality.

At present this is achieved by incorporating the chipsets from each of the different standards into the handset, e.g. 3G and BlueTooth. While such an approach works, it is relatively inflexible. An alternative is for communication devices to be designed like computers with general-purpose processing capabilities and different software for different applications. Such devices could then call up, or download, the appropriate software for the particular communications requirement currently in use.

The underlying architecture needed to achieve this is termed software-defined radio (SDR). In the future this flexibility might enable the more efficient use of the available spectrum through rapid deployment of the latest radio technologies.

While there are potential benefits of SDR there are many issues with its implementation. They include:

- Difficulties in implementing broadband antennas able to cover a wide spread of radio spectrum while retaining useable radiation patterns and small size.
- Lack of sufficient processing power. Although processing devices such as DSPs are becoming ever more powerful, the rise in complexity of mobile standards is faster, with the result that it is becoming ever harder to implement new standards in software.
- Insufficient battery power. DSPs and other general purpose circuitry are invariably more power hungry than circuitry optimised for a particular application. Battery capacity is only growing slowly.
- High cost, particularly in handsets. Given that the benefits of SDR do not generally flow directly to the user but are more relevant to the network operator, subsidy will likely be needed to introduce SDRs.

In practice, the benefits of SDR appear relatively minor compared to the issues. The current approach of multi-modal devices works well and will likely always be less expensive than a general-purpose SDR radio. Further, since new network technologies are generally introduced much less frequently than users replace handsets, there is little need for a handset to download a new standard when this can more readily be embedded in the handset during production. There may be more scope for software changes at the higher protocol layers (e.g. layer three and above) and there will certainly be substantial scope for application downloading. Because of this, we do not expect 'true' SDRs that can change their radio behaviour to be implemented during the next two decades. We do, however, expect handsets to be able to download a wide range of new applications and software patches.

### Smart Antennas

Smart antenna technology has the potential to significantly increase the efficient use of spectrum in wireless communication applications. Through intelligent control of the transmission and reception of signals, capacity and coverage in wireless networks could be improved.

Various smart antenna techniques may be used. These include:

- Antennas that form narrow beams which are steered towards the user. These result in a stronger signal received by the user and reduced interference to others. However, larger arrays of antennas are needed to form beams, and tracking moving users can be problematic.
- Multi-antenna diversity schemes (such as MIMO, described above). These have the potential to dramatically increase data rates, but in practice finding an environment with rich scattering but a slowly changing channel is difficult, and the cost of implementation is high.
- Semi-smart antenna schemes such as those that decrease the size of a cell with heavy loading and increase the size of neighbouring cells to compensate. This can be achieved with mechanisms such as antennas with variable downtilt. These schemes have lower potential gains than the ones described above but are cost-effective and simple to implement and do not require antenna arrays.

These approaches are complementary, some being most appropriate for large, costly infrastructure systems, others working best in certain propagation environments, such as where multipath is prevalent.

In the case of mobile handsets, smart antennas can potentially offer increased capacity. However, smart antenna deployment is complex and expensive for handsets and the business case for this application is uncertain at present. For the cellular operator, implementing smaller cells with conventional technology is both less expensive and less risky than implementing smart antennas. For the wireless local area network (LAN) application, where the size, power and processing complexity constraints are relaxed, access points with smart antenna technology have a stronger business case and indeed early versions are now available on the market.

A semi-smart approach, applicable to both mobile base stations and to wireless LAN access points, would be less complex and cheaper to introduce, and might provide some useful, though smaller, gains in spectral efficiency. Therefore, the business case is stronger and the technology could be introduced in the near term.

In summary, while smart antennas have strong potential, we do not expect to see them make a significant impact over the next 10–20 years. This is because most schemes are difficult to implement, do not always bring gains and require large arrays of antennas at base stations at a time when environmental concerns are high. Comparatively, installing smaller cells brings much greater capacity gains for a smaller cost and lower risk.

## Wireless Mesh Networking

A wireless mesh network utilises other users in the network as nodes to relay information. In this way information can be transmitted from one user to a distant user via multiple hops through the other users.

Mesh networks can be separated into fixed mesh, where the nodes do not move and new nodes are only added infrequently, and mobile mesh, where nodes move and frequently join or leave the mesh. Fixed mesh networks are much more effective than mobile because:

- It is simpler to establish mesh connections and less signalling is needed to keep them up to date.
- Directional antennas are simpler to deploy as some time can be taken to locate nearby nodes which will then remain static. These antennas can increase range and reduce interference.
- Fixed systems often use antennas mounted relatively high, perhaps at rooftop level, which increases the chance of finding other nearby nodes and establishing a mesh.

Many advantages are claimed for mesh networks, including:

- a limited need for infrastructure, reducing deployment and ongoing operational costs
- increased capacity because each node acts as a mini base station
- increased coverage due to the ability to hop around corners.

However, a significant body of research [6] casts doubts on these claims. Unless a node is connected via backhaul to the core network, it cannot effectively generate capacity. Instead, it needs to relay any communications that it receives. There are some gains due to the lower power that can be used for multiple short hops compared to one long hop, but these are outweighed by inefficiencies such as signalling protocols and the fact that the short hops are unlikely to align well with the single 'long hop' and so power requirements increase. The research showed that a mesh network would not provide more capacity than a conventional cellular system.

This situation improves as more nodes are provided with backhaul connections, reducing the number of hops any communication is likely to need. However, this then tends more towards a conventional cellular architecture.

Mesh systems can operate with less infrastructure than conventional systems, but the price paid for this will be lower capacity and likely longer delay as the packets of information hop from node to node. Such a trade-off may be viable in some situations but is unlikely to be sensible for mainstream wireless communications. Mesh systems can also act as a coverage extension for cellular systems, with mobiles just out of coverage having their signal relayed by another node. However, this does add substantial complexity and may not be reliable.

For fixed systems the situation is somewhat better. With directional antennas, increased spectral efficiency becomes possible. The coverage gains may also be important, especially at higher frequencies such as 28 GHz where line-of-sight propagation becomes critical.

Our view is that wireless mesh networks will not make a significant difference to high-volume wireless communications. Mesh networks may have niche applications such as working alongside existing networks to fill in areas of poor wireless coverage, in areas where conventional networks are uneconomic such as the provision of broadband services to rural communities, or in deploying sensor networks.

## Interference Cancellation

Interference cancellation (IC) is a technique whereby a receiver analyses the complete set of all signals it receives and attempts to remove those that are considered to be interference. Its operation is most readily understood at a CDMA base station. The base station will be receiving the signal from all the mobiles in the sector. It can extract the signal from a particular mobile using the correlation of CDMA codes. However, for weaker signals this may be difficult. The base station could decode the strongest signal and then subtract this decoded signal from the overall set of received signals. It could progressively do this, reducing the interference on the weaker signals until they can be decoded. As well as doing this sequentially, it could in principle be performed in parallel using an optimal detector, although in practice the complexity of these is normally far too high to be implemented.

There are many other situations where IC could be used. For example, a fixed link could have a separate antenna pointing at an interfering link. The signal from this antenna, suitably attenuated and phase-shifted, could be subtracted from the signal on the main link.

However, IC systems suffer from two problems: they tend not to work particularly well, and they are often costly. For IC to work, the interfering signal must be accurately characterised. If this does not occur, then the wrong signal may be subtracted, potentially worsening the error rate. The cost of the systems tends to be high due to the need for additional processing, in some cases additional antennas, and possibly the need to send additional information such as the codes in use.

Our view is that in most cases the cost and impracticalities of IC outweighs any potential benefits and we do not expect to see it widely used in wireless communications systems of the next 10–20 years.

**Cognitive Radio**

This was discussed in Chapter 4 where we concluded it would not have a significant effect on wireless communications.

## 6.4.2 Technologies Lowering Cost: Backhaul

Cells have to be connected back into the infrastructure. This could be via:

- cabling, perhaps copper or coaxial (if copper, a range of different types could be used from ADSL to dedicated leased lines)
- fibre optic cable
- fixed wireless, using microwave or free-space optical links
- mesh wireless, using a range of different approaches.

Each of these have different attributes as shown in Table 6.1.

The table suggests that there is no single ideal solution. If fibre or good quality cabling is available nearby this may form the best solution. If not, fixed wireless may be best, unless low data rate and long delays can be tolerated where mesh might be viable.

There appear to be no significant advances on the horizon for any of these technologies. Cabling would benefit most from ways of automating and simplifying trench digging – perhaps some kind of robot digger? Fixed wireless would benefit from mechanisms to avoid the need for line-of-sight signal paths, and mesh from approaches that can increase capacity and reduce delay. With the exception of potential advances in mesh, none of these seem likely. Even with mesh, there seem many good reasons why it is unlikely to be viable for most backhaul requirements.

**Table 6.1** Attributes of backhaul options.

| Technology | Data rate | Cost | Installation | Other issues |
|---|---|---|---|---|
| Cabling | From 1 Mbits/s for ADSL up to ~100 Mbits/s for leased lines | From $20/month for ADSL to ~$'000smonth for leased lines | Required. Trenches may need to be dug to bury cables. | |
| Fibre optics | Essentially unlimited | ~$'000s per month | As above | |
| Fixed wireless | Up to 1 Gbits/s for some solutions | In the region of $50 000 one-off fee | Simple. Units installed at both ends of the link | Spectrum needed |
| Mesh wireless | Up to ~10 Mbits/s at present, may rise | May only be ~$'00s | Very simple if mesh dense enough – just turn on unit | Plenty of potential problems with capacity and delay |

Hence, we do not expect to see any significant technological changes leading to reduced backhaul costs and availability. However, as increasingly dense fibre, copper and possibly fixed wireless networks are deployed in urban areas; this will ameliorate many of the issues listed above. For example, as fibre networks become more widely deployed, the additional distance needed to connect any particular cell site to the available fibre becomes progressively shorter, reducing costs.

---

**Fibre radio**

If fibre were widely available then some have suggested the deployment of a technology known as 'fibre radio' or 'direct radio'. This is a concept where, in its purest form, the base station antenna supplies its received signal to an electrical-to-optical converter. This is then connected to a fibre optic cable taking the signal back to a network node where the wanted signals are extracted and routed as appropriate. Transmission works in the converse manner.

Proponents claim a range of advantages for such a concept. In particular, each antenna would, in principle, be able to receive signals spanning a huge range of frequencies and equally could transmit across this range. As a result, almost 'the entire spectrum' would be available in each cell, allowing massive capacity. The antenna unit would also be very compact and cheap, potentially allowing widespread deployment of very high capacity, small cells. Some researchers have even considered passive

*(continued)*

electrical-to-optical converters such that the antenna unit would not require power. However, this does limit the range to around 10 m.

However, there are many problems with such an approach. First, such a device could not use the entire spectrum – some would remain reserved for a range of other applications such as satellite transmission and cellular systems. In the worst case, the device might not have available to it any more spectrum than a standard multi-band cellular base station. Second, while the antenna unit may be cheap, this approach places maximum requirements on the backhaul, which as we have just seen is more likely to be the bottleneck than the radio spectrum. Indeed, each antenna unit needs its own dedicated fibre connection going all the way back to the network node. If there were a dense deployment of antennas this might require a massive upgrade in fibre deployment. To understand this, consider that in laying fibre along the streets typically tens or maybe hundreds of strands of fibre are laid. Together these provide many thousands of gigabytes/s of capacity. But if each strand is dedicated to one antenna, and each antenna is providing coverage over say 100 m, then assuming only a few people in the coverage area at any moment, the total data rate being delivered may be well under 1 Gbyte/s. Essentially, the fibre is being used very inefficiently, so much so, that much denser fibre deployments will likely be needed. The cost of these is likely to outweigh any advantages at the antenna unit. Third, removing the electronics from near the antenna to some central network node brings few benefits – the cost of the electronics is low and falling over time. Finally, there is little need for a solution of this sort. A standard WiFi base station will provide enough capacity for the foreseeable future, is low cost and simple to deploy, and can be connected to a wide range of backhaul options.

So while we see fibre radio as an interesting architectural concept there does not seem to be sufficient drivers to overcome the substantial disadvantage of requiring dense fibre deployment.

### 6.4.3 Technologies Enhancing Interaction with Terminals

Handset features and services have become steadily more advanced over the last few years. However, many have noted that they are widely unused because the complexity of operating the service using a small handset is just too high. If wireless services are to become increasingly feature rich, then interaction with the terminal will need to improve. Interaction can be split into:

- *Data out* – broadly through the display. Large displays allow information to be presented more readily and make viewing experiences better. Touch-screens facilitate data input.
- *Data in* – this might be through enhanced keyboards, touch-screens or perhaps via speech recognition. The more options and the simpler they are, the more users will interact with terminals.

Each of these areas is now covered, through a contribution by an expert in the field.

**Displays**
*by Joel Pollack, CEO Clairvoyante*

As multimedia comes to portable devices such as cell phones and media players the need for increased luminance will become clear, so today's portable devices with a brightness of 200 candelas/m$^2$ (nits) of brightness will soon be specified to have 400 nits going to 600 nits. If one thinks about the LCD TVs today they are typically 600 nits and are in one's living room. Why wouldn't handheld devices need as much brightness to produce the same results for video and TV content? This is especially true if these are often used outdoors where the ambient brightness is much greater. The challenge will be to find ways to achieve this brightness while continuing to cut power consumption to keep the devices small and lightweight.

Certainly batteries will not see dramatic improvement in the next 15 years based upon the nature of improvements so far. We will likely see fuel cells become more popular and perhaps practical in 15 years, but that alone will not solve the energy problem for portable devices. I expect that technologies such as Clairvoyante's PenTile technology will take hold by then and will cut the consumption of power in half. PenTile technology is a technique that utilises a reconfigured layout of the subpixels in the display array, adding a clear (white) subpixel, mapping the traditional gamut to R, G, B and W and then performing subpixel rendering to enable a high-resolution display to be built with one-third fewer subpixels. Not only does this reduce the number of transistors in the backplane array, but it increases the open area, or aperture ratio, generally allowing as much as twice the light to pass through the display and thereby cutting the power of the display backlight in half (www.clairvoyante.com).

I predict that transflective (partially transmissive) and reflective LCDs will see significant improvement in the next 15 years. The ultimate portable display still looks like paper. Techniques will be found for transflective technology to make the whites better and to improve the colour gamut. Not only will viewability improve for apparent brightness, but the speed will also be quick enough to enable some level of video.

As GPS, web browsers, games and other multimedia permeate the mobile market we will see a trend to higher resolution displays, only limited by what the human eye can see (330 dots/inch). We will see VGA in 2.4-inch diagonal and XGA in 4-inch diagonal. The 4- to 6-inch XGA format will be used for ultra-portables which will detach from much larger notebook PCs to go to meetings with people to carry their presentations and calendars.

As people begin to use more data centric applications on phones and adopt ultra-portables there will be a trend to more digitisers for displays with improved handwriting recognition. Instead of digitisers behind the displays they will increasingly use the TFT array itself to capture information where the display transistors are used as phototransistors. Not only will they be able to recognise a stylus pen, but they will be used as image scanners, bar code readers and cameras.

White LEDs will predominate as backlights for larger and larger displays as they improve in luminous efficiency. They are far more convenient to use than cold-cathode

*(continued)*

fluorescent lamps (CCFLs) that have been used as the backlight source in notebook PCs, but are not easily dimmed. The ability to build TFT backplanes on to plastic will emerge within the next few years, and while it may be a little expensive at first it will eventually take hold for portable applications where durability and weight are at a premium. These plastic displays will remain rigid although lightweight. The truly flexible, roll-it-up, high-resolution display is probably further off than 15 years.

We will continue to see higher levels of integration monolithically on to the display backplane as carrier mobility of low-temperature polysilicon[1] (LTPS) thin-film transistor backplane arrays and its variants improve. One will see touch-panel controllers, graphics chips, speakers and audio amps, video algorithms and more finding their way on to the glass itself. Attached to the display will be drivers and other logic that is produced at ever smaller process geometry – 24 microns today moving to 15 microns and then to 10 microns. The effect will be slimmer and more powerful products.

The lifetime of organic light-emitting diodes (OLEDs) will continue to improve as techniques such as our PenTile technology reduce the demand of drive current. These, taken together with new adaptive techniques for backplanes, may make OLEDs useful for more of the portable video applications of the future.

Improvement in autostereoscopic displays (3D without the need to wear any glasses or other headgear) may happen in the next 15 years that remove some of the 'sweet spot' viewing issues, but still 3D will remain a niche that has limited acceptance.

For the TV space, LCD TV will predominate as improvements are made to speed, uniformity of off-axis gamma, and manufacturing cost. Within 15 years there will be a trend to more LED backlight versions where contrast and colour gamut is significantly improved to make images much more lifelike. These will migrate to the midrange market as cost of manufacturing continues to improve. Continued improvements in graphics chips will further reduce video noise and remaining artefacts seen in today's products.

---

[1] Behind the liquid crystal layer in higher performance, active matrix LCDs, there is an array of field-effect transistors that are addressed by row and column drivers. These are referred to as thin-film transistors (TFTs). Data is sent on the column drivers as each row is switched one by one, scanning from top to bottom of the panel. The multiplexing capability of the array is performed with the array of transistors, not with the liquid crystal material per se. There are two types of transistor arrays in common use today: amorphous silicon (aSi) or polysilicon. Polysilicons originally needed to be manufactured at high temperatures (> 600 degrees Celcius) where conventional glass melted, requiring the use of toques quartz (fused silica) to endure the processing temperature. Later people learned how to use annealing steps for a process that was able to work at < 600 degrees Celcius to use larger sheets of conventional soda-lime glass, or low-temperature polysilicon (LTPS). The advantage of polysilicon is that the transistors have higher carrier mobility such that electrons move more freely allowing higher data rates and performance unavailable with aSi TFTs. With such technology the drivers for the columns and rows could be fabricated at the same time as the transistor array and still allow for sufficient performance of shift registers and the like. Additional functionality for increasingly complex circuits are being added step by step to these monolithic (on-the-glass) ICs in the variants of LTPS being developed today.

Interacting with displays will see some change as the conventional mouse and trackball is replaced in some applications with field sensor search coils that can detect motion in more directions. They may be implemented in a glove that has multiple sensors and is as affordable as today's higher end mouse.

Micro displays are displays that are built on to silicon wafers with extremely high resolution. They can be used either for projection by reflecting light from their surface or as virtual imaging devices mounted into glasses or goggles. These will survive as more efficient optics is developed allowing these products to become increasingly important to retail advertising and other consumer front and rear projection applications.

I would like to point out that nothing I have mentioned is really so new. The bases for all of these ideas exist today either in higher end products, R&D, or as a combination of new things today that will be combined in the next 15 years.

### Mobile Human/Computer Interaction
*by Tom MacTavish, Motorola*

Over the next 10 to 20 years, we foresee that mobile devices will become easier to use based on increasingly insightful design, the devices' resident capability and their seamless integration with systems around them. The most visible part of mobile devices, keypads and displays, will become increasingly convenient, adaptive and intelligent. Future mobile devices will assume many forms to become an integral part of tools used by productive society. Three trends supporting these conclusions are emerging.

First, there is a general movement towards increased sophistication in user interface capabilities that is based on consistent increases in the computational power and memory sizes available for small, low-power form factors. This impacts both the physical mechanisms people can use to communicate with mobile devices and also the underlying intelligence that allows those systems to accommodate various kinds of sensors, to host various interaction capabilities, and to adapt to different levels of user knowledge. Currently, we are seeing innovations that offer flexible direct text input mechanisms that provide choices to users who have touch-screen or input sensor enabled mobile devices. Figure 6.6 shows the type of virtual keypads available as touch-screen graphical overlays that are appearing to facilitate data entry tasks that could vary from roman characters, ideographic characters or numeric characters input.

By placing a capacitive sensor beneath a smooth surface, new keypads will permit direct data input with the human finger. Figure 6.7 illustrates a numeric keypad that also serves as a graphical character input mechanism. One could imagine this type of technology being further integrated into everyday consumer products, walls and furniture so that any kind of device capable of communication could receive input from a user.

*(continued)*

**Figure 6.6** Different types of virtual keypads: ideographic, character input and QWERTY keypad.

**Figure 6.7** Capacitive text input.

Further, the various tactile data entry techniques are being coupled with underlying intelligent systems that provide word and phrase prediction capabilities so that users will only have to begin writing a word or symbol, and based either on their statistical frequency of use, or in years to come, based on their intended meaning, the rest of the phrase will be predicted accurately. Why not leave a message for an absent neighbour by literally writing it on their door?

As speech recognition, natural language processing, synthetic speech, and interaction management capabilities increase their capabilities on mobile devices, we expect that people will be able to interact with devices at a more strategic level. Rather than requiring the user to know about a device's capabilities and invoke certain functions, we can expect that the device will be able to lead the user through tasks and suggest appropriate steps to a successful conclusion. This ability to lead the user has been demonstrated at modest levels by 'wizards' that help us install software in personal computers, but we can expect much more robust capabilities in the future. A more natural interaction based on language means that mobile devices of the future may find new physical forms that are more wearable, or conformable to the person or the tasks at hand. If a microphone and speaker are the only requirements to access and control computer-based intelligence, then industrial designers can create new classes of devices with few constraints. The most advanced manifestation of this trend will result in virtual conversational agents and more sophisticated devices supporting physical robot/human interaction. An early precursor to these new-type devices is suggested by MIT's Media Lab 'Kismet' project where they are conducting interesting research on using animatronics for novel presence detection and communication tasks.

Second, we observe two trends that relate to system architectures and the development processes used to build user interfaces. Industry standards are emerging that support a greater amount of interaction between multiple devices, for example, allowing one appliance to be controlled through a user interface on another. This means that we will likely see two device embodiments.

- Those that act as our universal interpreter of devices so that we can walk into a room and have the device discover other compatible devices and provide us with familiar control mechanisms. If I want to adjust the volume on an unfamiliar audio device that is in earshot, I may have my mobile device discover the logical controls for the audio device, present me with familiar volume controls, and allow me to raise or lower its volume as needed.
- The second type of device-to-device interaction may by driven by a device that can 'borrow' resources as needed to support the user's immediate task at hand. For example, if I wish to share multimedia content available on my mobile device with my colleagues standing nearby, why not allow me to stream the video and audio feed directly to a nearby large-screen TV?

Third, we note a number of trends that relate to user lifestyles and behaviours. Personal expression has become a much more important application of computer and communication systems, particularly among young people. Also, as time has become a more and more precious commodity, technology is allowing traditional boundaries in time and place between our different activities to become more blurred and technology has responded by finding ways to make the blurring more manageable. With the advent of personal video recorders, consumers are able already to shift the time for viewing, from the time of its broadcast, to a time of their own convenience. Also, they are able to create and broadcast their own content and create their own unique audiences and communities. And, perhaps as a result of these new abilities, media viewing is becoming both a more social and a more interactive event. Those behaviours related to generating, consuming and managing content at a personal level will engender new kinds of devices. We can foresee a world in which people have their own personal, mobile projector so that they

(*continued*)

can share their self-generated or unique community content instantly with their friends. Or, they can use this new kind of mobile projector to enjoy larger format media and richer media without having to find specific time or places specially outfitted for 'movies'. Also, we are observing a gradual increase in the visibility and acceptability of wearable technologies that allow communication and computing tasks to integrate more seamlessly into our everyday lives. It is no longer an oddity when someone wears an ear-microphone arrangement through the grocery store checkout lane, or when we hear someone talking loudly and interactively while alone in a public thoroughfare.

When considered in full, these major trends suggest that the mobile devices of the future will be intelligent, ubiquitous, adaptable, predictive, pervasive and chameleon-like extensions to our own abilities. Maybe we won't see them at all because they will be part of our existing wardrobe, appliances or tools. If my device knows me well and can assist me as needed, then we will have created transparent entities that increase my power without burdening me with new requirements and operational demands.

## Speech Recognition
### by William Meisel: President, TMA Associates
### (www.tmaa.com)

Wireless technology enables mobility, but smaller devices limit the ways in which we can take advantage of that mobility, particularly as the number of services and features proliferate. Small screens, difficulty in entering text, and the distractions of other activities (e.g. driving) encourage the use of speech technologies such as speech recognition. Speech recognition and its inverse (text-to-speech synthesis) can allow interaction strictly through voice dialogue.

Is speech technology up to the task? To answer this question, we need to understand a platform distinction. Speech technology in the network works just as it does when you phone your bank's call centre from any phone: nothing is processed in the phone. The speech is transmitted to a server in the network and processed there. 'Embedded' speech technology, on the other hand, can operate in the device itself, but is limited in its capabilities by the memory and computational limitations of the device. Voice dialling by speaking a name is a typical application of embedded speech recognition, and is a common feature on wireless phones.

Network-based speech recognition and text-to-speech have matured and are now reliable and flexible. Getting an account number by voice is now considered an easy task. Companies are even using speech recognition to automate changes-of-address and to provide directory assistance listings. Text-to-speech synthesis can now sound like a recording, with only minor hints that it is synthesised. Synthesis allows one to listen to emails or to hear information from text-based databases. If you tried speech recognition in a call centre and didn't like it, you may have simply encountered a bad design—the technology shouldn't be blamed for a designer's bad decisions (such as preventing you from talking to a customer service representative).

The capabilities of embedded technology have steadily grown. It is now possible to dictate a text message or email using software in the phone. Since embedded technology doesn't require the wireless connection be active, such applications can allow composing a message or performing other tasks without incurring connection charges or when a connection isn't available. Embedded text-to-speech software is available to speak the information on a mobile-device screen for those who have difficulty seeing it. It hasn't been done at this writing, but embedded technology could be combined with network-based technology to provide a near seamless experience. From the point of view of users, they talk and the device replies; embedded and network-based speech technologies appear the same.

One advantage of network-based speech technology is that it can be reached from any phone and – in effect – creates a voice user interface on that phone no matter what the phone characteristics. Network-based applications also typically support using the touch-tone pad which can provide a fallback solution when the environment is too noisy for speech. Another advantage is that there is a well-accepted standard called VoiceXML for developing speech and touch-tone applications. (VoiceXML is maintained by the World Wide Web Consortium, the same group that manages many web standards.)

Network-based speech recognition will likely predominate over embedded recognition. Servers in the network will always have more processing power than handsets, and perhaps more importantly, will not be limited by battery life, as handsets are. Network-based systems can also more readily be updated with new names and terms, such as the name of a new artist offering ring tones for download. Embedded recognition might still be used for accessing device functionality, such as selecting a music track on an MP3 player.

Speech technology can work with other modes of interaction beyond the keypad. For example, one can request information by voice and have it delivered as text to the phone. This 'multimodal' approach is particularly useful if one wants to retain the information or if it would take too long to listen to it. VoiceXML supports modes other than audio and touch-tone, particularly playing video clips on devices and networks that can support video, and is being extended to support additional user interface modes.

Speech technology is up to the task of providing an effective voice user interface for mobile devices. History has shown the importance of the user interface in the acceptance of a new technology. The graphical user interface (GUI) – implemented by windows, pointing devices, icons and menus – led to the adoption of the PC and web browsers by the general population and took them out of the realm of the hobbyist and university researcher.

Service providers and entrepreneurs are trying mightily to get customers to use advanced wireless services, with limited success. Perhaps it is not the services that are at fault, but the difficulty of using them. Speech is the one modality that is constant among almost all wireless devices. It is the prime candidate for helping wireless devices meet their potential as multi-purpose tools and ubiquitous electronic assistants.

We now summarise and draw out the implications of these thoughts:

- Regarding displays, without foldable or roll-away screens in widespread use, mobile devices will generally have similar form factors to today. However, the displays are likely to be brighter, higher resolution, have touch-screen facilities and possibly be able to scan barcodes and similar. Many of the current constraints, such as the difficulty in viewing TV on a small screen, will remain over the period of our forecast.
- Concerning form factors, initially we will see advances in capacitive surfaces that will allow people to draw on their device or other surface with their finger. Then devices will become cleverer, working out through context and history what the user is trying to do and presenting appropriate interfaces. Finally, the user's device will interact with other nearby devices presenting control mechanisms in the preferred manner. Over time devices will integrate into clothing or other personal items to become increasingly invisible.
- Speech recognition will advance, mainly through network-based recognition, relatively quickly to the point that speech interaction becomes the preferred way of interacting with devices. However, there will be a distinct difference of capabilities according to whether the device is connected to the network or not.

So overall the ability to interact with terminals will improve somewhat, through enhanced user interfaces, better speech recognition and slowly advancing enhancements to displays. However, battery life will hamper developments, impeding brighter displays and better processing such that interacting with the mobile phone will progress slowly. In the longer term, interaction will improve through increasing device intelligence and the ability of a device to control other electronics in the vicinity. Around 20 years from now we might expect devices to 'disappear' as they are integrated into clothing and interaction becomes entirely speech-based and intelligent.

### 6.4.4 Technologies Leading to 'Artificial Intelligence'

Many visions of the future, including those put forward in our earlier book, assume that the phone will become a 'remote control on life', finding coffee shops, setting daily alarms and monitoring the house and car for problems. Very little of this has happened yet. Indeed, very little of this type of intelligence has occurred anywhere. Perhaps early examples are on the PC, with Windows pre-populating certain fields on web forms such as 'name' and modifying the menu structure on some programs depending on usage. Most likely the reason for the limited evolution is the complexity of the problem. Even working out when to set a morning alarm given a travel schedule is very difficult unless the user has entered all the information in a very rigid fashion and has already set within the device parameters like the time taken to leave the house in the morning. However, as with all problems of complexity, this is one that can likely be solved with sufficient processing power.

The computing power to achieve this need not be in the handset. Instead, all of these programs could be run within a network, and simple instructions, such as 'set alarm for 7:00', be sent to the handset. Indeed, there may be other advantages of having the function located in the network, such as allowing the user to change handset and enabling simpler connectivity to other associated entities such as corporate networks. In the network there can be plenty of processing power. Hence, it is more a case of the time needed to develop appropriate software and gradually extend it across various areas of user behaviour. Even this will likely take decades.

### 6.4.5 Compression Technologies

When audio and video signals are converted to digital they are typically compressed in order to reduce the bandwidth required for their transmission. As compression technology improves, coding rates have fallen, allowing more users to be accommodated within a given bandwidth.

Voice coding rates fell steadily during the 1980s and early 1990s but now seem to have stabilised in the region of 10 kbits/s for a reasonably high quality. On current trends, substantial improvements in this seem unlikely.

Video coding rates, however, have been halving around every five years. For example, H.263 requires approximately half the bandwidth of H.261, and MPEG4 (H.264) only half that of H.263. However, it is far from clear whether this trend can continue for any significant time. Although there are many ideas for high compression systems these typically require very high processing powers which either cannot be achieved, or if they can will result in too high a battery drain for mobile devices. Our view is that, like voice coding, video coding will soon level off at a rate not readily improved upon.

Hence, we do not expect changes in compression technology to have a significant effect in the next 20 years.

## 6.5 Technology Prognosis: No Key Breakthroughs

Summarising the chapter so far:

- The technical efficiency of wireless systems is approaching fundamental limits. While there is some prospect that MIMO antennas might increase capacity beyond these, this prospect seems relatively slim.
- The many empirical laws suggest that capacity and data rates on wireless networks will continue to increase but primarily driven by ever smaller cells.
- There are no technologies currently under investigation that promise any dramatic change in the technical efficiency of wireless systems.
- Display technology on mobile phones will steadily improve and network-based speech recognition will bring significant gains. Interacting with mobile devices will improve as devices become more intelligent and learn user behaviour, but interacting with the mobile will remain compromised compared to devices such as PCs.
- Steady progress is expected in areas such as batteries, processing power etc.

## 6.6 Implications for the Future

***Technical enhancements will not bring about change directly***

All these conclusions suggest that there is no wonderful technology, or technical trend, that on its own is going to revolutionise wireless over the next decade, and with all likelihood over the next 20 years. Equally, we can expect increased capacity and data rates almost entirely as a result of smaller cells. These might be a mix of cellular networks continually reducing cell sizes in urban areas, and the increasing deployment of W-LANs, in some cases providing coverage across entire cities.

As cell sizes get smaller, many of the technologies, such as MIMO, become less viable. In the case of MIMO the smaller cells have less multipath and so the MIMO benefits will be reduced, while the costs of MIMO will remain similar to those of a large cell, making it less attractive economically. Mesh systems get pushed 'to the edge' as ubiquitous small cell coverage is built. But this is not problematic – small cells with current technology can readily provide massive increases in data rates and capacity compared to current cellular systems. For example, a ubiquitous W-LAN network could provide data rates in excess of 1 Mbits/s or higher depending on cell density.

However, different problems emerge. The key cost element for small cells, particularly those offering high data rates, soon becomes the 'backhaul' – that is the connection of the cell into the core network. Other costs include site rental and power but for small cells in urban environments these costs are typically low. While backhaul of any required data rate can be provided through the deployment of appropriate copper or fibre cabling, this can be uneconomic for small cells serving only a few customers. Indeed, we can go as far as to say that since small cells are the key route to increased wireless data rates and capacity, and since backhaul is the key constraining factor in the deployment of small cells, that the biggest challenge facing wireless is fixed communications.

As discussed above, advances in backhaul seem unlikely. Instead, steadily increasing backhaul density will reduce costs, driving smaller cells, which in turn will drive increased backhaul. This virtuous circle will help lead the way to ever smaller cells, but because of the time needed for wireless to drive wired deployments and vice versa, it will slow the process of moving to cities covered with microcell sites.

# References

[1] C E Shannon, 'A mathematic theory of communications', *Bell Systems Technical Journal*, vol. 27, pp 379–423, July 1948; and vol. 27, pp 623–656, October 1948.

[2] C E Shannon, 'Communications in the presence of noise', *Proc. IRE*, vol. 37, pp 10–21, January 1949.

[3] W T Webb, 'Modulation methods for PCNs', *IEEE Communication Magazine*, vol. 30, no. 12, pp 90–95, December 1992.

[4] W T Webb, 'Spectrum efficiency of multilevel modulation schemes in mobile radio communications', *IEEE Trans. on Communications*, vol. 43, no. 8, pp 2344–2349, August 1995.

[5] W C Y Lee, 'Spectrum efficiency in cellular', *IEEE Trans on Vehicular Technology*, vol. 38, no. 2, pp 69–75, May 1989.

[6] See http://www.ofcom.org.uk/research/technology/overview/emer_tech/mesh/

# 7

# Major World Events

## 7.1 Introduction

So far in this book we have concentrated on issues and events within the wireless industry. However, it may be that the industry is significantly shaped by events outside of its control. Broadly these would be 'major world events' that would have an impact on many aspects of our lives. For example, the sort of events that might be foreseen in 2006 include:

- global warming accelerates
- fuel crisis
- major conflict in the Middle East
- terrorism increases substantially
- global recession – pensions crisis
- China becomes dominant
- home working and remote working grows dramatically, perhaps as a result of one of the above.

In addition, there may be events in related industries that could have an impact. For example, on-line and multi-player gaming might experience massive growth.

The purpose of this chapter is to explore the possible consequences of some of these events on mobile communications. While it is difficult to predict whether and when any of them will happen, understanding their implications provides some degree of sensitivity analysis.

## 7.2 World Events

### Global warming accelerates

The result of this might be an increase in severe weather conditions and perhaps a rise in sea level leading to increased flooding. In the disaster areas there would be potential damage

*Wireless Communications: The Future*   William Webb
© 2007 John Wiley & Sons, Ltd

to all kinds of telecommunications networks. This might lead to a desire for more resilient communications systems. For example, mesh capabilities might be built into cellular handsets so that if the network collapsed at least an ad-hoc network could be formed. Overall, though, unless the disasters were very widespread, the effect on telecommunications systems would be small.

Another impact of this, and many other of the scenarios described below, would be a diversion of funds, away from consumption and towards rebuilding and strengthening. This would ultimately result in individuals having less disposable income. This might reduce the growth in spending on communications; indeed in extreme situations it might cause it to fall. However, in a world becoming more uncertain, many would see wireless communications as increasingly important and would likely increase the priority of spending. Overall, then, the financial impact on wireless communications may be relatively limited.

In summary, we would not anticipate this scenario causing significant change to the predictions set out in later chapters.

### Fuel crisis
In this scenario oil becomes scarce, or very expensive, or both. This might be triggered by some conflict, or perhaps by a dramatic rise in consumption in China. Alternatively, terrorist activity or global disasters might affect large parts of the supply industry. There might be two effects of note – people travel less and the disruption triggers a major global recession. The latter is discussed in more detail below. The former might likely increase the need for communications. This might lead to more home working and greater use of conference calls. Both might act to spur the introduction of video telephony. The availability of broadband connections will also become important, perhaps stimulating increased fibre optic deployment, although any recession might militate against significant investment.

In summary, we would expect this scenario to speed the use of communications, perhaps predominantly fixed, but there might be an increase in wireless usage too.

### Major conflict in the Middle East
This might lead to increased fuel prices and global instability. Both have been covered in earlier scenarios.

### Terrorism increases substantially
The effects of this might be to deter people from travelling, to increase financial instability and to cause localised disasters. Some of these have been discussed above – less travelling will increase the need for communications, and localised disasters will make people more reliant on wireless communications. Financial instability is discussed in more detail below.

Overall, this outcome would likely result in increased communications and more rapid deployment of new systems.

### Global recession
Many events could trigger a global recession. Some might result in a short-lived drop in activity, perhaps akin to the Internet bubble bursting and causing a recession from 2001 to around 2004 in the communications industry. Others might be more long term – for example

a pensions crisis might take a generation to resolve, dampening activity for decades. Some of the impacts of a recession might be:

- Consumers have less disposable income, although as discussed above it seems likely that other discretionary expenditure will be reduced before spending on wireless communications.
- Operators might be less inclined to build dense networks, or deploy new ones. However, our predictions are for limited new network deployment in wireless in any case. It is possible that 3G deployment might be slowed.
- Manufacturers might reduce research and development activity. In particular, this might slow the speed of handset developments.
- Consolidation of operators might occur sooner to reduce competition and increase revenue. Potentially, this might result in the 'pipe' versus 'service provider' arrangement coming about more quickly than would otherwise occur.

Overall, we would expect the effect on wireless communications to be minimal, partly because consumers will continue to spend and partly because we are not predicting major investments in the future in any case.

### China becomes dominant
Some predict that China will become a key player in wireless, shaping the industry, dominating manufacturing of equipment and becoming the largest market for terminals in the world. To date, China's involvement in wireless has been somewhat disruptive because of their tendency to aim for local standards such as TD-SCDMA. Our prediction is that this disruptiveness will diminish as China becomes increasingly dominant and hence has less need to influence standards in order to grow its industry.

If this were to happen, it seems unlikely there would be many effects on the wireless industry. Some non-Chinese manufacturers might shrink, to be replaced by Chinese companies, but this would have little impact on the overall future direction.

### Home working and remote working grows dramatically, perhaps as a result of one of the above
This was already discussed as part of earlier scenarios.

## 7.3  Events in Related Industries

It is harder to predict these without a detailed knowledge of other industries. Overall our view is that we cannot conceive of a 'disruptive technology' or 'new entrant' that removes the need for wireless communications – there are no known alternatives! The only events that we could imagine would be those that increased the need for wireless communications or related services such as location information. For example, if multi-player gaming were to become much more popular, users might play while mobile, increasing the wireless traffic levels and stimulating development of more advanced terminals. Hence, it seems much more likely that related events will have only a positive impact on wireless communications.

## 7.4 Summary

We have considered a range of major events. We certainly hope that few if any of these befall us in the next 20 years. Our discussion and summary from each of the events showed that in most cases wireless communications will be unaffected. In some events, it may even gain increased investment, or more rapid acceptance of new services. In others, it is possible that a recession might delay some investment, but our view is that this will not have a significant impact on our future. Broadly, short of a catastrophic breakdown in society, our view is that wireless communications will be more affected by the issues we have been considering in earlier chapters, rather than major world events.

## 7.5 The Next Chapters

The next six chapters provide a range of views from experts across the field of wireless communications. They extend from visionaries to CTOs, from operators to manufacturers and consultants, from those working in military spheres to those working in commercial areas and academia. We then bring together all their views in Chapter 14.

# 8

# Future Military Wireless Solutions

Paul S. Cannon and Clive R. Harding

## 8.1 Introduction

Communications via radio, cable or optical fibre pervades almost all aspects of military operations. For the distribution of commands it provides person-to-person and one-to-many connectivity. Communication systems also provide the vital conduit between one computer and another to ensure the distribution of logistical information, intelligence and data from sensors. As such, a communications capability is integral and crucial for the operation of many military systems that do not have explicit communications functionality; for example, data retrieval from space-based surveillance sensors and command and control of unmanned air vehicles (UAVs). Communications systems can be viewed as the glue that cements military operations and, given that modern military operations are highly mobile, much of this communications is carried out wirelessly.

In order to ensure information dominance, military communications must be maintained when and where they are needed. They should be resistant to jamming, direction finding and other electronic warfare (EW) threats. They must also provide end-to-end message security in order to maintain privacy. The communications network must be robust to physical disruption – in time of war, an architecture employing a single critical communications node, such as a computer or radio station, has to be avoided if possible. Further, military communications often have to be maintained on the move, for example between fast-moving aircraft, command posts afloat and moving troops onshore. This may have to be achieved without a fixed or preinstalled infrastructure.

Fully meeting these requirements significantly constrains radio system architectures and information flow rates relative to those that can be achieved in civilian systems. For these various reasons R&D into military communications remains critical. As government is often an early adopter of new technology, these new military communications technologies can sometimes provide an important stimulant to commercial civilian wireless technology.

*Wireless Communications: The Future*   William Webb
© 2007 John Wiley & Sons, Ltd

## 8.2 Operational Context

In the not so distant past, military communications were required to support air, land and sea battles against a well-known and well-defined enemy and it is inconceivable that such requirements will disappear within the next 20 years. However, recent experience and changes in the geopolitical backdrop have added new and testing military communications requirements – namely truly worldwide operations and operations in an urban environment. To this has been added peace-keeping and operations other than war (OTW), where military forces aid the civil authorities in emergency and disaster relief.

Many military communications systems still in service had their genesis during the cold war. Then requirements were for communications systems capable of supporting specialist roles in well-defined regions: e.g. anti-submarine operations in the North Atlantic, and combined land and air defensive battles on the North German plains. A high priority for systems was their ability to operate in an environment that might be subject to nuclear, biological and chemical (NBC) attack, as well as being highly resilient against deliberate electronic attack and exploitation. The extant doctrine, types of operation and supporting equipment did not require the passage of large quantities of data to large numbers of users drawn from disparate coalition forces. Other technical features, such as the ability to withstand a nuclear electromagnetic pulse (EMP), were unlike those required in the then more limited civil domain. These various and unique technical requirements led to specialist manufacturers developing and supplying bespoke military communications systems. Such systems were developed over many years, with the expectation of a long in-service life (measured in decades).

With the ending of the Cold War, and rising international terrorism and instabilities, the nature of military operations has now changed to one where there is far greater uncertainty over where forces will be deployed or what role they will be fulfilling. Rather than mainly being equipped to fight in a regionalised high-intensity conflict against a similarly equipped enemy, military doctrine now recognises that forces may be engaged simultaneously in operations across the spectrum of conflict (peace-keeping, conflict, post-conflict reconstitution) and humanitarian support, in a variety of theatres worldwide. These operations are likely to be conducted as part of a coalition of other nations, as well as in conjunction with different host nations.

Communications in dense urban environments is now an important military requirement. In this context, secure, small, lightweight, cost-effective, high-bandwidth communications systems are needed to allow high-capacity short-range communications between various platforms and between individual combatants. A patrol involved in building-to-building clearance of a dense city area could use such a system to exchange tactical views (from gunsight or helmet cameras). In addition, images from UAVs, drones or robots could be relayed to the combatants to improve their situational awareness.

The numbers of military personnel in most nations have also decreased, with a compensating increase in the technology supporting them (sometimes described as 'fighting smarter'). Concepts such as 'network centric operations' (NCOs) seek to facilitate the seamless movement of information across a variety of (previously stove-pipe) bearer media, to enable the conduct of complex operations at a high tempo (see Figure 8.1). Such concepts require radically different communications from those employed during the Cold War.

At the core of NCOs are communications networks that must be as reliable, available and survivable as the platforms to which they connect. In contrast to Cold War communications

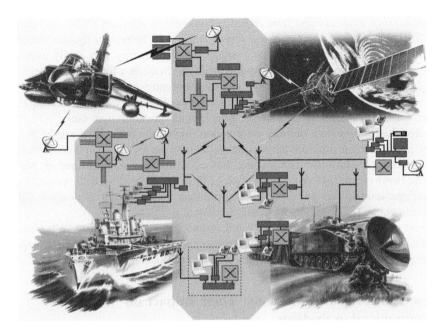

**Figure 8.1**   Network centric operations [reproduced by permission of © QinetiQ].

these new networks must distribute huge amounts of data quickly (from a plethora of sensors) across a wide area. This is no simple requirement and it will require a number of radical changes in the communications systems. To some extent this may be brought about by algorithmic developments but radical and developmental change in technology is likely to be a catalyst. R&D will be required in areas that include mobile ad-hoc self-forming networks, information assurance and security, spectrum management, heterogeneous networks and low-vulnerability communications more generally.

NCOs are the backdrop against which the next 20 years of military communications development will occur. That said, the introduction of military equipment is often painfully slow. We expect that military communications procurements cycles must and will shorten. At a minimum, fielded functionality must be refreshed and upgraded through technology insertion programmes in response to new and emerging requirements and threats.

## 8.3 Technical Features Important to Secure and Robust Global Military Communications

Of fundamental concern to the military is the reliability and security of their communications. Military operations would be seriously jeopardised if these essential services were lost or compromised in some way. The problem is made worse by the strain placed on the communications systems operating in hostile environments, particularly given sparse and often overloaded resources and a determined enemy capable of EW and other methods of attack.

Techniques such as spread-spectrum code division multiple access (CDMA) and frequency hopping (FH) have been used for many years. These help to protect against interception and jamming attack by spreading the radio frequency (RF) energy very thinly over a wide bandwidth or by hopping quickly and randomly from one frequency to the next. The objective is to be below the noise threshold at the enemy interceptor or, to give them insufficient time to search and lock on before skipping to the next frequency. These techniques, first developed for military communications, are now in widespread civilian use.

To protect highly sensitive military information, high-assurance cryptography is widely used. The protection this provides aims to ensure that the enemy takes an unrealistic amount of time to recover useful information using the best available computers and cryptanalytic techniques. However, the effective application of this type of cryptographic protection adds significant operational costs and complexity to military communications systems. Effective key management and distribution are vitally important.

For some operations, such as disaster relief, support from the military is often essential. Inter-working with relief agencies and civil emergency communications services is important, and interoperability, based on common standards and protocols, is imperative.

## 8.4 New Platforms and Missions: Their Impact on Military Communication Systems

### 8.4.1 Impact of Unmanned Vehicles

Unmanned vehicles (UVs) have long been considered as alternatives to manned platforms for dull, dirty and dangerous operations. UVs offer advantages in terms of endurance, expendability, survivability and cost. Consequently their use is set to expand with the number and types of operation increasing over the next two decades. UVs are classified according to the domain in which they operate and are most commonly referred to as unmanned air vehicles (UAVs), unmanned ground vehicles (UGVs), unmanned surface vehicles (USVs) and unmanned underwater vehicles (UUVs).

A primary role of many UVs is as a sensor platform supporting intelligence, surveillance and reconnaissance (ISR) operations. Reliable, high-capacity communications links, capable of carrying high-quality video imagery, will be a critical element of these systems. Near-term developments will see the UVs physically returning sensor information for later exploitation. However, rapid exploitation of UV-mounted sensors requires some means of transmitting the data back to the military commander, avoiding the need for the return of physical storage media. The transmission of imagery requires high data rates, particularly where multi-spectral sensors are involved. Data compression (e.g. compression, auto target recognition) will relieve this requirement to some extent, at the cost of additional processing and reduced robustness.

The use of man-portable UAVs and UGVs, particularly in urban operations, is also set to expand. The generally smaller size of UAVs in comparison with manned aircraft constrains the size, weight and power consumption of communications equipment. This is particularly evident in the case of micro-UAVs, which have very limited power and nowhere to site a high-gain antenna. Thus, for this class of UV high data rates over anything but the shortest ranges will be unachievable.

The provision of communications on UAVs designed for low observability (LO) will be a challenge if the platform characteristics are not to be compromised. A major challenge is the provision of antennas with good communications performance, yet also having low radar and infrared signatures. Plasma antennas have been assessed in this context.

UUV missions, such as mine countermeasure and anti-submarine operations, will continue to be dominated by acoustic communications. However, developments in specialised antenna technologies will see the introduction of short-range RF systems. Primary use of this technology will be the elimination of connectors for high data rate communications over very short (less than one metre) ranges. Lower frequency variants will offer low data rate communications at ranges up to 1 km.

### 8.4.2 Impact of High-Altitude Platforms (HAPs)

High-altitude platforms (HAPs) have been the subject of intense study in recent years for both civilian and military uses. Fundamentally a HAP offers the possibilities of an almost geostationary and long-term hook in the sky at about 21-km altitude. A HAP at this altitude will have a communications footprint on the ground of about 200-km diameter and visibility of about five million cubic kilometres of airspace (see Figure 8.2).

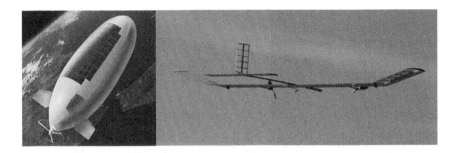

**Figure 8.2** (left to right) Artist's impression of a lighter-than-air HAP; and QinetiQ's 'Zephyr' [reproduced by permission of © QinetiQ].

A platform having these parameters opens up many new operational possibilities and is stimulating many teams to explore their benefits in, for example, communications, remote sensing, air-space management and surveillance. The most influential exponent of the HAP is the USA. Here programmes sponsored by DoD and NASA are pushing the technologies towards new military and commercial capabilities. However, we are also seeing significant developments in the EU, Japan, Korea, China and elsewhere [1, 2].

A HAP is both a UAV and a low-altitude pseudo-geostationary satellite. A conservative comparison of link budgets for HAPs and a geostationary (GEO) communications satellite gives about a 40 dB improvement in the uplink and a similar improvement for the downlink (all other parameters being equal). A HAP can, therefore, sustain more traffic or access smaller terminals than a GEO. An obvious use of a HAP would be to support all of the in-theatre communications with satellite communication (satcom) reach-back to national headquarters. A network of HAPs could extend the footprint to cover larger theatres with inter-HAP links and a GEO command and control overlay.

Geostationary military satcom in the UK has grown at a staggering 50% year-on-year since the late 1960s (see Figure 8.3). Most of this growth has been as a consequence of user demand and NCOs will exacerbate this. We believe that the military use of HAPS to support satcom is inevitable within the next 20 years and probably much sooner.

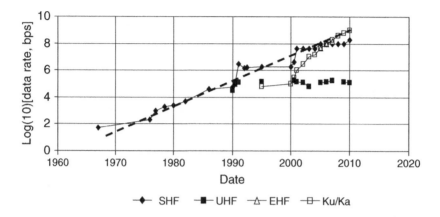

**Figure 8.3**  Growth in UK military satcoms since the late 1960s. The dashed line is for 50% year-on-year growth and matches these data quite closely [courtesy of Dr Alan C. Smith, reproduced by permission of © QinetiQ].

## 8.4.3 Impact of Future Infantry Soldier Technology

The advent of the digitised battlespace has lead to an increase in the complexity of the electronic equipment and associated wiring harnesses used by infantry soldiers. This trend is likely to continue with the development of advanced infantry concepts that involve sophisticated helmet-mounted displays and video rifle sights. Wired connections between all such subsystems make the infantry soldier's personal equipment cumbersome to use and can limit the effective exchange of information between soldiers. These problems can be ameliorated by the use of wireless connections between electronic subsystems and the development of effective body-area networks. Key to the delivery of this capability will be low-power solutions together with innovations in battery and primary power technology.

## 8.4.4 Impact of Wireless Sensor Networks

As has been discussed earlier, information superiority is of central import on the battlefield. A significant element is the acquisition of physical information about the location and disposition of enemy forces. Typically, this is gathered using high-value platforms taking, for example, photographic and video footage of geographic areas of significance. These platforms tend to be expensive resources that need careful management. Further, the loss of one of these platforms seriously degrades the intelligence gathering capability.

A new generation of low-cost, disposable sensors promises an alternative to this. A network of relatively simple sensors, coupled with auto-configuring communications and networking technologies, can provide a cost-effective method of gathering information about the environment, and actors in that environment. The development of low-cost sensors is driving research into arbitrary array formation (sensor and communications) and associated resource management to maximise power, bandwidth and sensing capability.

Defence funded research is starting to tease apart the complex interactions and the trade-offs inherent in wireless sensor networks. Initiatives such as the Smart Dust programme running at University of Berkeley, California, and funded by DARPA are attempting to push tiny sensors to the limit with target package sizes of a cubic millimetre. Whilst this programme has its sights on the devices of the future, the next ten years will see a wide variety of military sensing applications met by networking broadly brick-sized and smaller devices.

## 8.5 Developments in Military Communications Systems

### 8.5.1 Introduction

The various communications systems available to the US military are being drawn together by the Global Information Grid (GIG) – in the UK this is known as the Global Information Infrastructure (GII). These represent the primary technical frameworks required to support NCOs and will remain so for the foreseeable future. The GIG is defined [3] as the globally interconnected end-to-end set of information capabilities, associated processes and personnel, for collecting, processing, storing, disseminating and managing information on demand to warfighters, policymakers and support personnel. The GIG encompasses all owned and leased communications, computing systems and services, software (including applications), system data, security services and other associated services necessary to achieve information superiority. The intention is that all advanced weapons platforms, sensor systems and command and control centres will eventually be linked via the GIG.

It is recognised that delivering an effective GIG will demand significant enhancement of the available technology and it is expected that the majority of the GIG capability will be achieved through exploitation and appropriate 'hardening' of commercial technologies and standards as well as bespoke military communications systems. In the remainder of this section we address a number of military communications domains and describe both their current status and how they may develop over the next 20 years.

### 8.5.2 Very Low-Frequency (VLF) Communications

VLF has largely fallen out of civilian use. However, the military find its ability to penetrate the sea, its high availability and its resilience against exo-atmospheric nuclear attack essential for the provision of broadcast command and control information to submarines. This is unlikely to change for the next 20 years.

Current standards are under revision with a view to increasing data rates from 200 bits/s to 800 bit/s or higher through a combination of more efficient modulation techniques, message coding and compression schemes. The drivers behind these developments are a need for increased range, a decrease in transmitter power and an increase in data rates.

## 8.5.3 High-Frequency (HF) Communications

HF, via both skywave and surface wave, is an important communications asset for the military. In addition to its strategic use, it provides the primary means for long-distance communications for many air platforms. It also provides a tactical communications capability via near-vertical incidence skywave (NVIS) and surface wave.

New data waveform standards, automatic link establishment (ALE) techniques and data transfer standards (ARQ) have been brought into operational use in the last five years. In tandem many countries are overhauling their HF infrastructure to include these new standards. These various changes have automated HF communications frequency management and have provided both high data rate (9600 bits/s) communications and robust lower rate communications when the ionospheric path is disturbed. As a consequence there has been a resurgence of HF usage.

The future development of HF can be broadly separated into the provision of higher data rates, developing and improving the data handling techniques (e.g. HF-based IP) and improving the efficiency (i.e. lowering bit rate) of existing digitised services.

Existing HF standards can provide data rates up to 9600 and 19 200 bits/s using channel bandwidths of 3 kHz and 6 kHz, respectively. Research has shown that further increases in data rate require either larger overall bandwidths (i.e. beyond 6 kHz) or the use of multiple independent 3 kHz channels. Current research is concentrating on the latter approach as it complies with existing international HF spectrum allocation regulations. This approach has yielded data rates up to 64000 bits/s skywave, albeit with the use of multiple HF equipments.

A number of development programmes aim to provide IP networking over HF by protocol enhancements to the media-access control mechanism of existing standards. We also note that the data compression and coding capabilities of MPEG-4 and JPEG 2000 and high rate waveforms have recently been combined to provide rudimentary videoconferencing using a standard 3 kHz channel. In the future we expect that further developments will enable the transfer of operational military reconnaissance data via these techniques. Commercial HF techniques are being investigated, in particular Digital Radio Mondale (DRM). However, there are issues with the robustness of the waveform within some military environments.

Rather than focusing on the provision of higher overall data rates, some recent areas of innovation have investigated data rate reductions for digital voice. The advantage of reducing data rates at HF is the subsequent increase in robustness, especially on disturbed ionospheric channels. Current standards provide relatively high quality digital speech at data rates of 1200 and 2400 bits/s. Following recent work a workable rate of 600 bits/s is to be added to the standard, providing an appreciable increase in robustness with a small corresponding reduction in voice quality.

Some interesting work is addressing the concept of multiple-input/multiple-output (MIMO) configurations but it is not yet clear whether this will result in any practical advantages. The multipath inherent in HF communications would superficially appear to offer an ideal medium for MIMO. However, recent theoretical modelling indicates that, as lower frequencies will be needed to achieve this, the resultant rays will be largely reflected from the E-region and will be highly correlated. This combined with the greater absorption experienced at these lower frequencies may offset any MIMO processing gains. Further study is required.

Given the relatively recent upgrades, HF military capabilities are expected to remain broadly static over the next 10–15 years, although there is flexibility in the underlying HF hardware to support technology enhancements.

## 8.5.4 Terrestrial VHF, UHF and SHF Tactical Communications

Owing to the nature of the military environment, military land-based communications systems differ significantly from their civil counterparts. From the system architecture point of view, almost all civil communications systems involve a fixed infrastructure that can be carefully engineered to operate well. Point-to-point links can be engineered to provide high bandwidth operation over paths at high frequencies and over long distances. Mobile systems can operate via a fixed base station, again sited and engineered to provide the required coverage. This approach means that one end of the link can use an elevated antenna with gain. Also the base station can control access and traffic routing. By contrast, in military tactical systems, if system engineering is possible, it is limited to what can be achieved by a limited staff in minutes. The options to deploy equipment are generally limited by operational considerations. Reliance on a single base station is generally unacceptable on grounds of vulnerability, although some long-range systems use this architecture with the base station out of theatre.

For military applications, VHF provides propagation over ranges suitable for combat radio applications. At present combat radio supports good-quality voice communications within all informed nets. Data can also be supported on this architecture. Typical deployments involve radios from different nets being co-sited in the same vehicle or field headquarters. This provides an opportunity for data to be exchanged between nets, so as to provide end-to-end connectivity spanning several nets. In the future, it is to be expected that VHF system architectures will change, in response to the desire for improved resilience and increased spectral efficiency, exploiting SDR and cognitive radio techniques.

However, the data capacity of a VHF link is limited, so the requirements of NCOs give rise to a need for a tactical system to support higher capacity data. The need for high data rates leads to the use of a higher frequency, which will provide a high data rate over a relatively short range. Again, the solution lies in a network across which the data can be transferred. For such a system to be effective, some degree of self organisation is essential. Systems of this type are already available. It can be anticipated that the techniques will develop further, so that self-organising, high data rate and spectrally efficient tactical systems become a major element of military communications.

There can be little doubt that tactical trunk networks, employing point-to-point links and directional antennas, will continue to provide superior communications data throughput and spectral efficiency where operational considerations permit their use. The use of UHF links will remain important, in this context. However, the desire for higher data rates is likely to cause continued pressure for the development of techniques which can exploit SHF. Troposcatter links will also remain important to provide long-range, high-bandwidth terrestrial links. It is difficult to say whether this technology will be replaced by satellite or HAPs.

Broadcast topologies – one transmitter to many receivers – will be essential for the distribution of large data products and situational awareness messages. Currently this is provided by satellite communications and HF, but the need for higher data rates and a requirement for more efficient use of spectrum will lead to the use of high- and medium-altitude relay platforms on UAVs, balloons or even HAPs.

Short-range voice communications are needed to provide contact with other members of a unit – avoiding the need for shouting and hand signals. There is also an emerging requirement for data communication, to allow pictorial, location and other non-voice information to be transferred. We expect that personal role radios (PRR) supporting this functionality will become increasingly important as part of an integrated approach to enhance the capability of infantry soldiers.

Frequency supportability and the associated spectrum management are likely to have a strong influence on system design. In particular, pressure for bandwidth in the UHF area of the spectrum, from both military and civil systems, may dictate that some systems must operate in other bands, with the consequent design implications. Pressure for spectral resource is likely to lead to the need for greater coordination of spectrum usage between systems as well as within the systems themselves. This is discussed further below.

### 8.5.5 Satellite Communications

Satellite communications (satcoms) are essential to support all aspects of modern military operations. They provide secure and flexible communications for maritime, air and land forces.

In the 1970s when satellite communications had matured sufficiently to be used by the military, it was generally viewed as the beyond line-of-sight (BLOS) communications fallback to HF communications for carrying low-speed telegraphic signals between shore and Royal Navy ships. By the early 1980s the situation had reversed, with SHF satcom routinely providing primary communications links between deployed ships and UK headquarters. Even the modest data rates provided (typically 2400 bits/s to 64 000 bits/s) gave UK based commanders the potential to be in direct contact with their forces, actively supporting the development of modern communications and control doctrine (see Figure 8.4).

**Figure 8.4** Military satellite communications providing in-theatre communications [reproduced by permission of © QinetiQ].

This use of satcom by all areas of the armed services has increased with the launch of military satellites carrying UHF and SHF payloads accessible by a wide range of ground terminals. The latter vary in size from small man-packs (supporting tactical voice), to medium size transportable terminals, with an antenna diameter of 1–2 m, through to 8 m fixed station terminals.

Military satcom requires high availability, survivability and a low probability of exploitation (LPE). High availability dictates multiple satellites to provide global coverage, with resilient links capable of operating in severe environments. High-priority communications links need to survive a wide range of potential degradation which may take the form of unintentional interference or sophisticated jamming. Consequently, military satellites avoid out-of-theatre interference by incorporating focused beams able to point to specific regions of the Earth's surface. Advanced modulation and error-correction techniques, together with the use of spread-spectrum transmissions, are then employed to make the jammer's job more difficult. To minimise the likelihood of unwanted detection or location finding, military satcom ground terminals use high-gain antennas with controlled sidelobes and spread-spectrum waveforms. Information security is further ensured through encryption.

One of the biggest challenges over recent years has been to make optimum use of available satellite channel bandwidth to support packet switched (e.g. IP) and circuit switched user services (e.g. dialled voice). This has been achieved by adopting novel gateway devices which enable standard protocols such as TCP, which were not originally designed for use over high-latency satellite links, to be re-encapsulated in protocols specifically tailored for such channels. These techniques enable important leveraging of commercial telecommunications technology (e.g. routers and multiplexers) to give improved satcom performance. A further benefit of adopting commercial standards is the added efficiency offered by the application of powerful data compression (up to 40 times) and voice codec (up to 10 times) algorithms. These and other adaptations of emerging commercial telecommunications techniques offer greatest promise for improved capacity and usability of military satcom over the next five to 10 years.

The cornerstone of UK military satcom is the Skynet 5 programme. This programme will deliver military satellite communication services to the UK Armed Forces until 2020. Skynet 5 will supply managed end-to-end services to include: switched voice, packet switched data and broadcasts, delivered to a range of availability and survivability levels. It will upgrade ship-borne and land-based terminals underpinned by new fully managed telecommunications equipment based on ruggedised COTS equipment. The solution will also use satellite links provided by commercial operators (e.g. INMARSAT) to supplement military satellites; in particular supporting welfare emails and voice calls home, critical in maintaining staff morale. Skynet 5 satellites will employ flexible phased array technology to provide steerable, variable beam profiles to optimise performance and offer users higher survivable capacity. Skynet 5 is a key enabler to NCOs providing fully managed guaranteed quality of service (QoS) BLOS communications.

The continued quest for jamming-resistant, LPE-secure satcom links has led to the use of higher frequencies which inherently support narrow beams from relatively small antennas, whilst offering sufficient bandwidth for wideband frequency hopped waveforms. Systems employing EHF (typically 44 GHz) uplinks and SHF (typically 20 GHz) downlinks are becoming available for deployment in a variety of mobile platforms and transportable units.

In the UK, a MoD programme called Naval EHF/SHF Terminal (NEST) aims to provide wide-bandwidth survivable satcom to the Royal Navy. NEST provides dual-band terminals, able to operate at SHF (up- and downlink) via Skynet 5 satellites, or EHF uplink, SHF down via the US Advanced EHF (AEHF) satellites. Skynet 5 satellites will utilise transparent transponders, often referred to as a 'bent pipe', because the signal from and to the earth terminals remains at RF and is simply amplified and retransmitted.

AEHF satellites will employ on-board processing. They will be based on an enhanced version of the 1990s era US Milstar constellation and will be able to support links up to 8 Mbits/s. Satellites with on-board processing demodulate the uplink signals on one beam and re-modulate these on to a new carrier, perhaps on another beam, for retransmission. This provides extremely high flexibility and jam resistance. Global coverage is achieved via a constellation of three AEHF satellites interconnected via broadband microwave cross links.

The next major performance leap for military satcom will occur through the US Transformational Satellite Communications System (TSAT) which will provide further enhancement of AEHF satellites to include integrated Internet like networking functionality. Five TSAT satellites are likely to be launched from 2015 providing a wideband survivable network centric capability to service the GIG in support of strategic and tactical warfighters. TSAT will employ packet switching with bulk and packet encryption/decryption to support secure information dissemination. TSAT's IP routing will connect thousands of users through networks rather than limited point-to-point connections. Additionally, TSAT will enable high data rate connections to space and airborne intelligence, surveillance, and reconnaissance platforms. Initially TSAT satellites will use traditional microwave cross links to AEHF satellites to achieve integration and support transition from circuit switched to packet switched service delivery; eventually TSAT satellites are intended to be interconnected by highly secure wideband laser cross links. Jam-resistant laser links are also planned to link TSAT to an Advanced Polar Satellite System which will provide secure communications with LPE strategic and tactical users in northern polar regions. TSAT offers enormous increases in total user bandwidth capacity with up to 2 Gbits/s per space vehicle compared with 250 Mbits/s for AEHF. User RF data rates up to 45 Mbits/s are planned, with future laser communication links promising 10–100 Gbits/s.

The USA also plans to upgrade its tactical UHF military satcom, replacing the present UHF Follow On (UFO) system with the Mobile User Objective System (MUOS) satellite. Delivery of MUOS space segment is expected to begin around 2010, with full implementation around 2015. UHF service will begin to transfer from the UFO system as soon as elements of MUOS become available.

MUOS will use the 3G mobile telephone technology W-CDMA standard which describes a direct-sequence spread-spectrum system, having a bandwidth of 5 MHz, to provide 64 kbit/s. This contrasts with the frequency division multiple access (FDMA) 25-kHz channelisation used in the UFO system. The standard will also be ported into ground terminals forming the Joint Tactical Radio System (JTRS).

In the 2015 to 2020 time frame, integrated satellite systems (geostationary and polar), using a combination of RF links (satellite to ground/ships/air) and laser communications (satellite to satellite and satellite to aircraft/UAV) links, will fulfil the primary BLOS military bearer requirements for reliable wideband, global secure connectivity, in support of network centric military operations.

## 8.6 Emerging Communications Techniques

### 8.6.1 Introduction

The development of the above communications capabilities is being underpinned by innovative developments in a number of areas. Some of this work has taken place to support civilian communications systems but much of it is specific to the military.

### 8.6.2 Ad-hoc Networks

Self-organising, or ad-hoc networks have no central control and no dependence upon fixed infrastructure. Military networks, by their very nature, must be mobile, robust to attack and resilient to failure. Ad-hoc networks, therefore, are a natural fit for many defence applications. There are an increasing number of planned military systems that, at their core, depend upon an ad-hoc network. These include command and control networks, sensor networks, smart munitions networks, swarms of robotic vehicles for land, sea and air, and personal area networks.

Recent work in developing underpinning theory for ad-hoc networks has raised substantial concerns regarding their scalability. A notable example is the work of Gupta and Kumar [4], who produced an influential paper estimating the capacity of large (in terms of the numbers of users and devices) ad-hoc networks given certain assumptions regarding, for instance, traffic patterns. This work concludes that the capacity available for each node in an ad-hoc network tends rapidly towards zero as the number of nodes increases. This is due primarily to the need for nodes to share their use of the spectrum with other nodes in their local neighbourhood. Quantitatively, Gupta and Kumar found that the available capacity per node is inversely proportional to the square-root of the number of nodes.

Civil networks for the foreseeable future will, by and large, circumvent the scalability concerns of ad-hoc networks by deploying them as small to middling stub networks on the side of a fixed infrastructure. (Possibly with the notable exception of wireless sensor networks that, whilst being large in terms of the numbers of nodes, will probably have low actual utilisation.) Whilst the military can mirror this architectural thinking to a certain extent, reversionary modes of operation must still be supported. Further, quasi-static trunk or satellite communications will struggle to provide a stable backbone to support highly dynamic, tactical operations. Thus, military ad-hoc networks will need to scale to larger numbers of devices, supporting a more dynamic mix of services than their civil counterparts.

Work to establish a 'fundamental theory' of wireless ad-hoc networks, including methods for secure operation of these, will mature over the next five years. Meanwhile developments in protocols and algorithms will increase the capability of ad-hoc networks towards the capacity bounds set by theory.

### 8.6.3 Disruption-Tolerant Networks (DTN)

Offensive techniques for deliberately attacking and disrupting information systems are becoming widespread; they are countered by information assurance techniques, with countermeasures put in place to prevent an enemy from compromising critical information infrastructure and systems. With the move towards NCOs by armed forces worldwide, the ability to disrupt such enemy information systems whilst remaining impervious to similar attack becomes increasingly critical.

Current offensive technologies concentrate their attacks on the wireless bearers (e.g. jamming) or the end computer systems (e.g. hacking, spoofing and denial of service). However, with the increasing move to intelligent IP-based packet data networks, the network itself – routers, switches, bearers, protocols – becomes a key new battleground.

Ad-hoc networking has been developed to extend to highly mobile devices and users the communications capabilities commonly enjoyed by fixed or cellular users. However, these networks are still dependent upon reasonably stable paths through the network. More specifically, the time to deliver packets through the network must be markedly less than the rate at which that path changes due to, for instance, mobility. If no stable path through the network is available when communication is required, conventional protocols such as TCP/IP break down.

Disruption-tolerant networks offer the promise of useable and useful communications across networks which are frequently disconnected with unreliable links. It has grown from work developing protocols for inter-planetary networks. These networks are subject to very long but, on the whole, predictable delays. Whilst this is arguably true of some military networks (e.g. UAVs in a patrol), many applications will have unpredictable disconnections. This can arise due to mobility, hostile enemy action (physical and electronic) or environmental conditions (weather, urban clutter).

To support communications in these types of network, DTN uses store and forward protocols. A fundamental component of information delivery in a DTN is the concept of bundles. Bundles are aggregated packages of user information that have common end-points and delivery requirements. Bundles can be passed between DTN nodes, repackaged along the way, as appropriate, until the information reaches the destination.

Data mules are another component of DTNs. These physically transport bundles between physical locations. An example of this is a helicopter making regular trips between ships in a fleet. As well as transporting physical cargo and personnel, the helicopter could also transport data. In this way DTNs can provide increases in both availability and capacity. These issues are illustrated in Figure 8.5.

The hexagons are network nodes, S being the information source, D being the destination and the rest intermediate DTN nodes. The left chart illustrates useful connectivity (the bars) when using conventional IP protocols. This is limited to just the time when a whole end-to-end path is available across all the nodes. Compare this to the right-hand chart. Here DTN protocols allow nodes to utilise links on the path between source and destination as they become available.

Experimental components of DTNs are available now for research. Some similar technology has been used in existing civil networks. A notable example is the provision of information to rural communities in the developing world [5]. DTN will form a useful tool for military networks, but it will be a number of years before operational deployment is practical.

## 8.6.4 Software-Defined Radio (SDR)

The military interest in SDR is driven by the need for integrated, flexible and affordable communications and networking capabilities. Particularly with the changing role of the

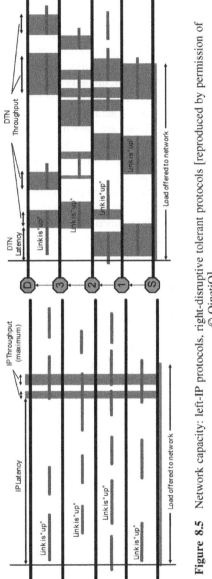

**Figure 8.5** Network capacity: left-IP protocols, right-disruptive tolerant protocols [reproduced by permission of © QinetiQ].

military there is a need to work, and thus interoperate with, a wide variety of allies. These needs will require incremental improvements in communications capability.

The military has traditionally been provided with radios with specific capabilities, procured to meet a specific operational requirement. As such it is often uneconomic or technically infeasible to upgrade existing radios with a new capability so new radios have to be procured. These either replace existing systems or, in many cases, need to be carried in addition to existing systems. This is clearly not ideal, especially in platforms such as aircraft where new equipment integration costs are very high and where weight and volume is a major consideration.

In the civilian world, SDR technology is recognised as an important driver of communications systems development and it is likely to have a similar impact on military communications systems. Military SDRs will include similar functionality and derive many of the same benefits as civilian communications but there will be military specific imperatives such as the inclusion of transmission security (TRANSEC, e.g. frequency hopping) and communications security (COMSEC, e.g. encryption). Military SDRs are also likely to lead the way in integrating multiple waveforms to facilitate communications with legacy users and communications with coalition forces using other standards. Further, military SDRs will enable multi-band and cross-band operation. The ability for a single radio to operate in multiple bands is critical to supporting extant waveforms and providing wide military interoperability.

Perhaps the single key military benefit of SDRs, which differentiates them from previous technologies, is the ability to change their functionality in-service by downloading a new waveform application. New waveforms may (in principle) include the modulation scheme, communications protocols, cryptographic algorithms and possibly network level applications including cross-banding and gateways. This allows new communications functions to be introduced in-service without, in many cases, making any changes to the installed equipment hardware.

Whilst the military has led the development of software radio technology, it makes extensive use of civil-led technology developments (e.g. DSP, FPGAs, RF components, analogue-to-digital converters). The first military initiative which will see well-developed SDRs fielded is the US Joint Tactical Radio System (JTRS) programme [6]. JTRS is seen by the US as a networking technology to enable network centric operations, rather than as a communications replacement programme, and is a key enabler to fielding largely IP-based networked communications. In order to achieve this, the JTRS programme is developing a suite of wideband networking waveforms.

Over the next 20 years new military radio procurements will be increasingly fulfilled by radios which exhibit SDR capabilities (whether mandated or not). Further, we expect that new equipment procurements (in the 2015 delivery time frame) will be very largely fulfilled by reconfigurable SDRs.

### 8.6.5 Environmental Modelling for Communications Management

The propagation of radio signals constrains the range and performance of any radio system. The military require worldwide operation at any time of the day and year, with communications taking place through the ionosphere, rain, sleet, dust, fire, in cities, at sea and many other absorbing and refractive environments (see Figure 8.6) [7]. Military radio propagation and its interaction with the environment is thus all encompassing.

Absorption (signal loss), refraction (signal bending) and diffraction are just some of the important propagation effects that can occur for a variety of reasons. At different frequencies

**Figure 8.6** The wide range of influences and propagation environments applicable to military communication systems [reproduced by permission of © QinetiQ, except top left reproduced by permission of © ESA].

radio signals may be affected by passing round or through obstacles. Radio signals may also be absorbed as they pass through clouds or rain or they may be unexpectedly bent by the Earth's atmosphere, consequently covering unexpected distances or creating holes in coverage. For these signals we are interested in how the radio waves interact with weather phenomena. Other radio signals may be slowed down and bent by electrons in the ionosphere at altitudes above 100 km, and for these we need to study space weather, which describes variations of electrons and other charged particles in the Earth's near-space environment. These and many other complex interactions of the radio signal with the environment often fundamentally limit the performance of wireless systems.

Exciting new work is now being undertaken to support in-building deployment, since this is important for covert and other military operations. New and challenging research is also likely to be needed on data assimilation (see Figure 8.7) [8] of real-time environmental data into the military propagation models – some of this work draws on weather forecasting research and we expect to see the impact of this research over the next 5–10 years.

## 8.6.6 Spectrum Management and Utilisation

Military communications systems are making the transition from operating in well-established deployments within limited geographic regions (e.g. NATO forces only operating within Western Europe in ITU Region 1) to deployments anywhere in the world. With a broadly fixed deployment in a known region the available spectrum was well established and agreements defined the bands available to the military. Detailed frequency planning and assignments could then be carried out to optimise use of the spectrum and deconflict usage between different nation's forces and equipment. The host nations were generally expected to be compliant, especially if conflict actually broke out.

Deployments to different regions are likely to find different patterns of spectrum usage, with coalition partners competing for the same finite bandwidth. Operations in urban areas, in particular, are likely to be especially difficult, not only because of the RF propagation difficulties inherent in the complex urban terrain, but also because of the high density

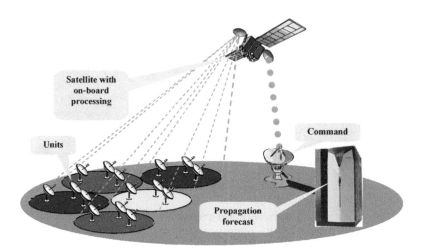

**Figure 8.7** Forecast driven proactive satellite resource management system [reproduced by permission of © R. Watson, University of Bath].

of civil spectrum usage and industrial RF pollution. This drives military communications systems towards being able to tune across a far wider band, as well as favouring technologies that allow wider bandwidths to be realised through the use of non-contiguous channels, or seamless routing across a disparate variety of bearers.

Given the foreseen increase in the number of systems using wireless communication and the drive for high communications bandwidths, effective frequency management is increasingly important. For many years, systems have been equipped with the capability to manage efficiently their own use of frequencies within the constraints of an allotment. With increasing numbers of different systems competing for the same spectral resources, the need for more detailed frequency management involving several systems is envisaged.

It is expected that, in order to maintain spectrum dominance, commanders will need to be able to evaluate and control spectral utilisation more readily than at present. An important aspect of military spectrum management is that of frequency assignment, which seeks to assign frequencies to different users such that they are free of mutual interference. In some cases, particularly combat radio, this leads to a need to assign, in an efficient manner, frequencies to hundreds of nets consisting of thousands of radios against a background of limited spectral availability. The result is a complex combinatorial problem. In order to obtain solutions close to the optimum, metaheuristic techniques such as simulated annealing and tabu search have recently been developed and tested. We expect to see further developments and applications in the next 5–10 years.

However, these methods give rise to a number of potential problems associated with the production and distribution of the frequency information. It would be highly advantageous for the network to be able to organise frequency utilisation autonomously. Whilst this represents a major challenge, the solution of this particular problem, using cognitive radio (CR) techniques, would show great benefit in the effective management of combat radio systems. CR has been defined by ITU Working Party 8 as 'a radio or system that senses, and is aware of, its operational environment and can dynamically and autonomously adjust its radio operating parameters accordingly'. This is a current area of research that provides promise of cooperative, autonomous, adaptive radio communications networks.

Civil CR is aimed primarily at use of 'white space' but cognitive radio should also address some of the military specific dimensions of radio use. Efficient or effective use of low probability of detection and interception waveforms, RF power management and smart antennas could help optimise spectrum usage. This wider use of the cognitive functionality needs to be examined in the coming years.

### 8.6.7 Smart Antennas for Military Communications

The pressure of spectrum availability for military applications, the need to increase communications capacity to support broadband applications and reducing radio component costs (driven by commercial personal communications developments) are making phased array antennas for communications increasingly practical. Smart antennas employ adaptive algorithms to optimise communications performance (e.g. by beamforming, improved jammer and interference suppression or advanced techniques such as MIMO). Multiple-input/multiple-output (MIMO) is a technique by which, in a multipath environment such as found in an urban environment or indoors, it is possible to exploit signal diversity to increase both communications robustness and capacity. Current areas of military research include applying MIMO technology to military communications where there is also a need to mitigate interference and jamming.

Whereas today smart antennas are limited to supporting niche military requirements (e.g. on high-value platforms), we predict that by 2020 smart antennas will be much more widely utilised in the military. We expect smart antennas to be in common use for broadband radio communications at UHF frequencies and above.

### 8.6.8 The Push to Higher RF Frequencies and Laser Communications

The push to higher radio frequencies makes available greater bandwidths and reductions in size – both attractive propositions for the military in much the same way as it is attractive to civilian users. The military can be expected to exploit the atmospheric windows, or conversely to achieve strong absorption in order to reduce the propagation range (reducing the probability of exploitation, or facilitating greater frequency reuse). The move to higher RF frequencies (60 GHz, 94 GHz) will be driven by the development of cost-effective semiconductor devices. Cheap 60 GHz devices are expected for the mass communications market in the next few years. Going beyond the traditional RF bands, 'free-space optical' (FSO) communication (also known as laser communication) systems may offer the military significant advantages.

FSO communication systems can, in principle, provide data rates that are far superior (e.g. gigabits/s) to those available on RF data links – but like millimetric systems weather (fog, turbulence) and some battlefield obscurants can degrade system performance and even render them inoperable. As such they are particularly beneficial when the communications path lies above the weather; i.e. aircraft-to-aircraft and satellite-to-satellite. The high data rates also mean that, even with data dropouts, burst mode operation can provide average data transfer rates which are operationally useful in the ground and naval context. FSO links provide some elements of security (such as low probability of intercept (LPI), low probability of jamming (LPJ)) as well as the potential for extremely low bit error rate (BER) characteristics.

The disadvantages include pointing, acquisition and tracking technologies, required to compensate for atmospheric scintillation, and platform motion, together with the effects of the weather, eye safety considerations and detection in a high scattering environment.

Whilst FSO technology has matured to some extent for fixed-site commercial deployments, the challenges faced with deploying FSO systems on mobile military platforms are still being overcome. The potential applications in the military domain are many, from short-range inter-ship links, to overcoming the data bottleneck for image transfers from air-to-ground platforms, UAV-to-ground troop data-dumps, air-to-air short-range links and inter-satellite or satellite-to-air links. It is expected that in the next 20 years the military adoption of FSO data links will increase.

### 8.6.9 Ultra Wideband (UWB) Techniques

Current conventional narrowband communication systems suffer from multi-path effects in dense urban environments which reduce radio performance. The bandwidth of these systems, and spectrum congestion, mean that relatively few high-bandwidth (video) channels can be accommodated in a given area.

One possible solution is ultra wideband (UWB), for which chip sets have been developed to take advantage of the US FCC spectrum allocation. The FCC has defined UWB systems to be at least 500 MHz wide in the band 3.1–10.6 GHz – the maximum permitted power level is low enough to allow UWB to coexist with others in the same band. Even so the UWB systems might provide $\sim$50 Mbits/s through buildings with a range of at least 20 m, as well as higher rates (up to 1 Gbits/s) at shorter distances. The low transmit power means the systems will have low power consumption and be small in volume. In addition, UWB transceivers can accurately measure their position relative to other transceivers and can act as short-range radar sensors. We predict that joint communications and position-aware UWB systems will be in military operation within ten years.

### 8.6.10 Communications Security

Secure communications networks are evolving in a number of areas, with NCOs a major driver. Secure end-to-end operation is required across a diverse range of military and commercial radio bearer and network technologies. Establishing secure sharing of information and associated trust relationships across multinational coalition network infrastructures is also required as part of developing concepts of operation.

Existing legacy systems make extensive use of cryptographic devices (cryptos) operating at the link layer but there is increasing application of cryptos at the network layer, particularly to support secure operation across IP-based networks. Effective application of network cryptography introduces a wide range of system integration issues in addition to those directly related to cryptographic protection. These include those relating to routing across different network domains, as well as operation across wired and wireless network segments with diverse link characteristics.

The implementation of high-assurance cryptography is also fundamentally changing with the adoption of programmable architectures that are able to support flexible operation using multiple cryptographic algorithms and interface types. Modules supporting programmable functions are already being utilised within cryptographic equipment as well as software-defined radios.

Operation of cryptos requires the routine provision of key material and this is generally supported by a dedicated key management system. Current military key management is largely based on the distribution of material for symmetric or shared key systems, but future system concepts are expected to make use of public key methods to negotiate the exchange of key material or enable flexible key distribution. This may include the provision of PKI (Public Key Infrastructure) or other techniques to support national and international needs for information security. The advent of programmable cryptos is also expected to require the distribution of cryptographic algorithms in addition to key material.

Over the next 5–10 years the increasing application of network cryptography will continue, especially in the context of converged packet and switched network services providing end-to-end voice and data services. Over the same time period programmable cryptos will be widely adopted in military systems. More widespread use of public key-based methods of key management is also predicted.

## 8.7 Some Emerging Technologies with Communications Relevance

### 8.7.1 Introduction

There are many technologies that may have a significant and possibly disruptive impact on the evolution of military communications systems development. In this section we introduce a small subset of these and indicate their potential significance.

### 8.7.2 Beyond Silicon Technologies

Silicon integrated circuit technology is developing along a road map roughly in accordance with 'Moore's law' – first articulated by Gordon Moore, the founder of Intel, who noted in 1965 that the number of transistors in an integrated circuit was doubling approximately every 18 months. Moore's law has so far held good for 40 years, and silicon semiconductor manufacturers recognise that they need to plan their development programmes to keep up with this trend, or else suffer a competitive disadvantage. Pundits have been predicting the eventual end of Moore's law for several years now, and yet technological advances have continued to keep progress on track.

It will not be possible to continue this exponential increase in device density indefinitely. The industry consensus, as presented in the 'International Technology Roadmap for Semiconductors', last updated in 2004, is that the current rate of improvement is sustainable until 2010, and may continue for several years after that, although progress may slow after 2010. The major challenges are the rising power dissipation of chips and the fall in power gain of smaller transistors. Economic factors are also significant; soaring fabrication plant costs are making it increasingly uncommercial to manufacture fully custom ICs that exploit the latest silicon miniaturisation. Finally, quantum effects in silicon will bring about an end to the ongoing miniaturisation of CMOS transistors in about 2018.

### 8.7.3 Potential of Nanotechnology

The behaviour of electronic devices at the nanoscale is radically different from that of conventional devices used in today's military and civil communications systems. This may provide opportunities for new generations of compact and very low-power electronic devices, which should bring significant enhancements to military wireless systems performance.

High-speed transistors have already been demonstrated and there is scope for development of other novel devices that exploit quantum resonances and other quantum phenomena. For example, spin interactions between a propagating electron in a semiconductor quantum wire or a carbon nanotube and an electron confined under an electrostatic gate or on a fullerene could be used for controlled spin manipulation and entanglement generation. Such devices have potential for application in quantum information processing elements, such as quantum repeaters for secure communications or, in the longer term, scalable spin-qubit arrays for quantum computing. The potential, within nano-electronics, of molecular electronics has recently been demonstrated. Two major advantages of molecular electronics are the potential higher operating temperatures (due to the higher energy scale of quantum confinement) and reproducibility. In the longer term, the cooperative behaviour of regular arrays of self-organised molecular networks could lead to completely new types of signal and data processing methodologies.

The main applications for nanotechnology devices over the next 20 years appears to be for very low-power electronics and ultra-sensitive detectors. Quantum devices are already leading to the possibility of new ways of communicating, with prototype quantum key-distribution systems and the longer-term promise of quantum digital information processing (quantum computing) with processing speeds for certain problems that would be unattainable with conventional (classical) computation methods.

## 8.7.4 Quantum Cryptography and Quantum Computing

Quantum computing, based around the superposition of quantum states, is highly speculative, and although no one yet knows how to make a useful-sized quantum computer such technology could be realised over the next 20 years. However, if it can be done, the massively increased computing power might crack some extremely difficult computing problems such as the rapid factoring of prime numbers to break public-key algorithms. This could then compromise military and commercial secure data exchange systems. Ironically, a solution to this potential threat may lie in quantum cryptography.

Quantum cryptography offers a method to support remote distribution of cryptographic key or other information. Unlike traditional cryptography, which employs various mathematical techniques to restrict eavesdroppers, quantum cryptography uses the phases of the individual photons or the entanglement of pairs of photons to ensure the secure distribution of the key material. The inherent security arises because unauthorised measurements leave a trace since the measurement disturbs the photon properties. In the case of the polarised photons this can be explained by Heisenberg's uncertainty principle.

Quantum cryptography is already emerging commercially as a technique that enables secure fibre or free-space optical links. In the near term, quantum cryptography offers potential as a secure distribution method for cryptographic keys but could also provide sufficient bandwidth for the distribution of random numbers to support more general information protection based on the use of 'one-time pad' cryptography.

Quantum cryptography, therefore, offers the potential for effective protection of information transfer from all known current and future forms of attack. In the context of military networks quantum cryptography has considerable potential to support secure information distribution, with operational applications that might include free-space optical links for cryptographic key distribution.

### 8.7.5 Negative Refractive Materials and Their Applications

In an ordinary material, the refractive index is a measure of the speed of electromagnetic waves through the material. This parameter is normally greater than unity, and the waves are slowed down in the material relative to free space. In contrast, negative refractive index materials cause the wave to increase in speed and this results in a number of novel properties [9, 10].

A considerable number of applications of negative index materials have been proposed including reduced size antennas and other RF components which could have a major impact on the design of radio systems. The process of testing the practicality of these devices against more conventional competition is still in progress. At the moment it is too early to tell if this new technology will be important. One-dimensional realisations of negative refractivity materials, such as circuits for dispersion compensation and filtering, may well be the first to find their way into applications.

### 8.7.6 Low-power High-stability Reference Sources

Oscillators play an essential role in all aspects of modern communications systems. At present only quartz crystal oscillators can be used in hand-held battery-operated devices because of their small size and power requirements. Atomic oscillators, the most stable oscillators developed, have until recently been too big to be used inside portable devices.

In August 2004, the US National Institute of Standards and Technology (NIST) announced the first successful demonstration of the inner workings of a chip-scale atomic clock (CSAC). These frequency references are expected to have an eventual size of $1 \text{ cm}^3$ and power requirement of $30 \text{ mW}$. This would be an improvement by a factor of 100 in size and power requirement over the current state of the art in compact atomic clocks. Although they are not as accurate and stable as normal size and commercially available caesium-based atomic clocks, they represent an improvement in frequency stability by a factor of over 1000 compared to a quartz crystal oscillator of comparable size and power requirements.

Miniature atomic clocks are expected to have a wide range of applications in military systems including spectrum efficiency. Perhaps the most obvious communications system benefit will be in the area of network synchronisation.

### 8.7.7 Power Sources

The US-DARPA Radioisotope Micropower Sources programme is expected to yield long-lasting, micro-scale power sources (RIMS) able to deliver 15–25 mW of continuous power with >40 mW bursts, within a size of under $1 \text{ cm}^3$ and a lifetime from one to 100 years (depending on the output power level).

Comparing this with current radio requirements, for example for Zigbee, RIMS are very nearly practical power sources for low-power radio. An example Zigbee device with a 75 m range has a power draw of around 30 mW while actively transmitting 250 kbits/s OQPSK at 0 dBm. It has a sleep power draw of only 1 μW. Today the peak power requirements are a little beyond RIMS, but not by much.

## 8.8 The Role for Commercial Off-the-shelf for Military Communications

There is no doubt that civil technologies are now leading military communications developments in terms of investment, performance and timescales. Enormous resources are available to develop cheap (mass market) equipment with increased functionality, to be used in a benign environment and with short lifetime expectations. Questions are, therefore, often asked of the form 'Why don't the military make more use of developments in civil communications?', often with the supplement 'surely the requirements are not that different and the functionality delivered by civil systems is superb – at a much lower cost?'.

Civil technology is indeed embedded in many military communication systems – this cannot be avoided if military systems are to deliver anything like the capability of their civil counterparts. Semiconductors, programming languages and operating systems are good examples. As we have seen, communications protocols and even cryptographic techniques often draw upon civil practice. Nevertheless, as we have also seen, the military requirements are sometimes very different from their civil counterparts. Therefore, the application of civil technologies must proceed intelligently and appropriately. In so doing the military designer is very aware that commercial off-the-shelf (COTS) technologies and products are readily available to allies and enemy alike.

Future military systems are expected to follow the development of COTS technologies with increasing convergence. Bespoke enhancements to civilian systems are a possibility; however, the increasing level of chip integration will likely make this approach more difficult as time progresses. All indications are that it is better to use COTS technology rather than COTS systems. Generally COTS systems will be insufficient on their own to meet military needs and targeted development will be necessary to address technology gaps and defence specific capabilities.

## 8.9 Summary and Conclusions

The work of Marconi, just over a century ago, started the wireless communications revolution. From its earliest days military applications drove the technology forward and this continued until the 1980s when the cellular revolution started. Since then the reverse has been true with technology transfer from the civilian domain into the military.

In the next 20 years we expect military communications systems to benefit from continuing and substantial investment in civilian communications systems and technology. We do, however, expect that there will be a number of areas where there will be important and exciting progress propelled by military research – especially those sponsored in the USA by DARPA. Some of these new initiatives have been identified above and no doubt there will be many more. The military is also an early adopter of new technology and this can stimulate the development of that technology for subsequent application in non-military applications. It is our belief, and hope, that in 20 years time it will be possible to look back on a period of more equal partnership between the civilian and military communications sectors. Whatever the case, we expect to see a huge increase in data throughput over military communications networks, driven in part by increasing use of adapted Internet protocols, the eventual realisation of optical and millimetric communications and band-sharing.

## Acknowledgements

The authors would like to acknowledge and give their grateful thanks to their colleagues who provided substantive contributions to this article: Ray Bradbeer, Ross Cuthbert, Nigel Davies, Kester Hughes, David Humphrey, Alan Smith, David Stephens and Paul Williams.

## References

[1] S Karapantazis and F-N Pavlidou, 'Broadband from heaven', *IEE Comms Engineer*, pp 18–23, 2004.

[2] D Grace, J Thornton, G Chen, G P White and T C Tozer, 'Improving system capacity of broadband services using multiple high altitude platforms', *IEEE Trans. on Wireless Communications*, vol. 4, pp 700–709, 2005.

[3] DoD, 'Directive 8100.01 Global Information Grid – Overarching Policy,' September 2002.

[4] P Gupta and P R Kumar, 'The capacity of wireless networks', *IEEE Trans. on Information Theory*, vol. 46, pp 388–404, 2000.

[5] A Pentland, R Fletcher and A Hasson, 'DakNet: rethinking connectivity in developing nations', *Computer*, vol. 37, pp 78–83, 2004.

[6] DOD, 'Joint Tactical Radio System website,' http://jtrs.army.mil, 2006.

[7] P S Cannon, J H Richter and P A Kossey, 'Real time specification of the battlespace environment and its effects on RF military systems', presented at Future Aerospace Technology in Service of the Alliance, Ecole Polytechnique, Palaiseau, Paris, 1997.

[8] M J Angling and P S Cannon, 'Assimilation of radio occultation measurements into background ionospheric models', *Radio Science*, vol. 39, pp RS1S08, 2004.

[9] D Sievenpiper, L Zhang, R F Jimenez Broas, N G Alexopoulos and E Yablonovitch, 'High impedance electromagnetic surfaces with a forbidden frequency band,' *IEEE Trans. on Microwave Theory and Techniques*, vol. 47, pp 2059–2074, 1999.

[10] R W Ziolkowski and A D Kipple, 'Application of double negative materials to increase the power radiated by electrically small antennas,' *IEEE Trans. on Antennas and Propagation*, vol. 51, pp 2626–2640, 2003.

## Biographies

### *Professor Paul S. Cannon, FREng, CEng, FIET, MAGU, PhD*

Paul Cannon is currently the Chief Scientist of Communications Division, QinetiQ. He is also a Senior Fellow at QinetiQ and pursues an active research career in collaboration with US DoD colleagues. He was formally Technical Director of the same Division and before that initiated, led and developed the Centre for Propagation and Atmospheric Research as a centre of excellence. He maintains a part-time appointment as Professor of Communication Systems and Atmospheric Sciences at the University of Bath, UK. Paul has published extensively with the results of his research being embodied in NATO communication systems. Paul has also sat on, and chaired, a number of NATO, national and international academic committees – he is currently Chair of Commission G of the International Union of Radio Science. Paul is a Fellow of the Royal Academy of Engineering.

***Clive R. Harding, BSc, CEng, FIET***

Clive Harding is currently Director of New Telecoms Ventures at QinetiQ. He has over 30 years' experience of military communications and the management of highly skilled professional teams dedicated to technical excellence. Clive has specialist knowledge of HF data communications systems and civil mobile telecommunications. He was formally Director of the Communications Department and most recently Chief Technology Officer for Quintel Technology, a QinetiQ telecommunications joint venture. Clive is a member of the Advisory Board for the Department of Communications Systems at Lancaster University and leads on the QinetiQ–Lancaster University partnering

initiative. He actively supports HF radio international collaboration through the IET international conference programme. Clive is a Chartered Engineer and Fellow of the Institution of Engineering and Technology (IET; formerly the IEE).

# 9

# From the Few to the Many: Macro to Micro

Peter Cochrane

## 9.1 In the Beginning

Early in the twentieth century, wireless sprang from investigations into electromagnetism with technology that now appears incredibly crude. Initial services were based on spark transmitters that occupied the full radio spectrum and interference was a major problem for inshore and offshore operations. But around 1915, DeForest invented the valve (or thermionic tube) and changed everything. He set in motion the direction and pace of radio development for the next 90 years. The ability to build amplifiers, modulators, filters, and precise and stable oscillators culminated in crystal control, followed by nuclear frequency standards later. All are still in use today, but mainly realised in solid-state electronics.

This all led to modes of analogue modulation (AM), frequency modulation (FM), single sideband (SSB) *et al.* for broadcast and public services that required channels and bands being allocated for specific purposes defined by power, noise, interference limitations, physical distance and service. For much of the twentieth century radio spectrum use was defined by analogue radio and TV broadcast services. Public radio for police, ambulance, fire and military quickly followed, and then the birth of the mobile phone saw a vast explosion of services with a gradual move from analogue to digital modes that were in themselves a variation on their analogue forebears.

All of this led to a dominant mindset that said: *The bands had to be decided internationally, the frequency channels had to be specified nationally and internationally, and modulation modes and coding schemes had to be fully defined, accepted and agreed.*

It is somewhat ironic that towards the end of World War II the desire to hide radio signals to avoid enemy detection and interception led to the development of spread-spectrum communication. This provided a coding scheme that is inherently opposed to all concepts

*Wireless Communications: The Future*   William Webb
© 2007 John Wiley & Sons, Ltd

of bands and channels and one that would ultimately be used for mobile speech and data services today in our 3G networks.

So herein lies a conundrum for the twenty-first century: *what to do with the illustrious history of radio that says bands and channels are the true way* The latest technology (born of the late 1940s) is now used in modern mobile networks, and it says ditch the past, just open up the radio spectrum and do what you like! Interference is no longer a big problem as our radios are increasingly intelligent and bands and channels are no longer required! How come? Sheer processing power not only allows spread-spectrum and sophisticated coding to be a reality, it provides us with the ability to negate interference by steering signals in frequency and physical space using active beam-forming and transmitter power adjustment (to be just sufficient) in a way that could not be envisaged 25 let alone 100 years ago.

## 9.2 The Need for Planning, Regulation and Control

The use and control of bands, channels, modulation mode and power has been almost entirely about ensuring an adequate communication path with minimum interference for all users (see Figure 9.1). By and large, almost all of our early systems operated in a broadcast mode, i.e. they sprayed radio energy everywhere. The only form of directivity available has been provided through the use of directive antennas that range from simple Yagis through to parabolic dishes in the microwave spectrum. But device nonlinearities, the

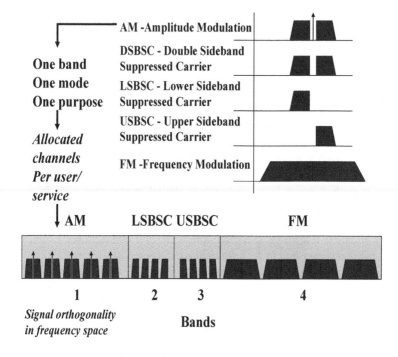

**Figure 9.1** Modulation and bands.

practical limitations of filters, proximity of transmitters and receivers, and signal scattering from static and moving objects, as well as antenna inefficiencies, has meant that interference has been a major engineering limitation.

Microelectronics and microprocessors give us the tools to alleviate the interference problem with sophisticated system designs that are far more tolerant of nonlinearities and relative power levels. Beyond the base design performance we can now employ complex coding and signal adaptation plus spatial beam-steering to realise additional degrees of freedom previously only available in theory.

Beyond all of this, advances in electronic devices allow us to operate at frequencies that only ten years ago were seen as impossible. For the most part, frequencies below 30 GHz have been allocated for use in just about every corner of human activity from broadcast to surveillance. Above 30 GHz, the allocation of channels is relatively sparse and indeed nonexistent if you go high enough. The reality is we have more radio spectrum available above 30 GHz than we have used hitherto. By and large, the spectrum above 90 GHz remains totally unused and can be extended to an increasing degree up to 300 GHz over the coming decades (see Figures 9.2 and 9.3).

In another curious twist of the technology story, the move to cellular operation for mobile devices and computers means that the localisation of radio signals is now the dominant mode. And the trend is towards smaller and smaller cells, often referred to as micro and pico cells that may span only 30 to 300 metres or so. Here then lies the route to infinite bandwidth and infinite reach for everyone. The signal bands above 60–90 GHz are ideally placed to allow us to communicate at an almost infinite rate from any device free of interference and devoid of any real scatter and signalling problems. And again, it is the availability of processing power that allows us to communicate black-box fashion without the need for complex user

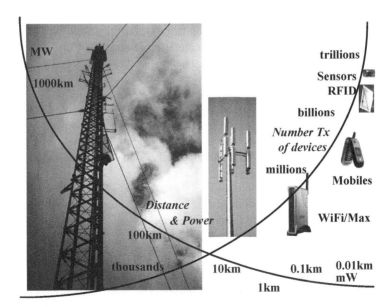

**Figure 9.2**   Migration from few to the many – towers to WiFi.

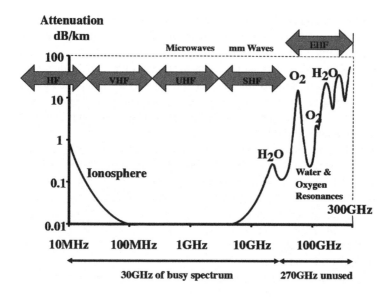

**Figure 9.3**  Wireless spectrum loss curve.

set-up routines or any degree of in-depth understanding of the technology, propagation or wireless operation.

In recent experiments, IBM have successfully clocked chips at 350 GHz at room temperature and 500 GHz when super-cooled. On this basis alone I think we might claim an existence theorem for an advancement of wireless working into the unplanned and unregulated frequencies above 100–200 GHz. We might therefore expect the use of spread-spectrum and ultra wideband systems providing data transfer rates well over 1 Gbits/s in the next 20–30 years, but over relatively short distances in the range 3–30 m.

These then are the likely key feature for the twenty-first century:

- A move away from bands and channels to spread spectrum.
- Increasingly sophisticated signal processing, coding and connection protocols with a high degree of automation that precludes the need for the user to get involved in any form of design, planning and set-up process. Just buy it, plug and play will be the dominant mode.
- Smaller and smaller communication cells with higher and higher bandwidths, connected into an optical fibre network that spans the globe.

All of this change is to some extent occurring of its own volition without the prior role of government and regulators other than to make the wireless space available.

## 9.3  Some General Trends

Over the past 50 years we have witnessed a revolution in the base technologies of wireless as well as the applications in the civil, military, fixed and mobile arenas. In order of importance the key core and peripheral advances/influencers can arguably be ranked as follows:

1 *Microelectronics* created new devices that took over from discrete valves and transistors to revolutionise system design. Big power-hungry radio systems (often) machined out of solid aluminium blocks have been replaced by very low-cost and small single/multiple chip-based solutions.

2 *Processing power and memory* have become commodities at an exponential rate with more intelligence and capability at the edge of networks than in the core. As a result, people can become net contributors and creators of services and data rather than mere users and absorbers of information.

3 *Mobility* has become a prime user mode and requirement for telephony and data with pocket and hand-held devices dominant in all networks. The mobile telephone revolution has created a new level of economic activity above and beyond all previous fixed communication networks.

4 *The PC* has empowered individuals by giving them tools and capabilities that were previously expensive, complex and the domain of a few specialists. This ubiquity of capability in the home and office has transformed productivity, companies, working lives and commerce. It has also been a prime mover in the globalisation of everything!

5 *The Internet* has created new information, trading, business and communication models, through the introduction of new freedoms, and has become a prime global business, education, social and commerce change mechanism.

6 *Complex control, coding and signalling* has improved channel capacities and resilience to interference. Simply controlling transmitter power in real time, and using crude sectorised (directional) antennas, huge advantages have been gained over earlier systems that were dumb! The addition of active signal agility in time, frequency and space through smart antenna systems will accelerate this feature even further. And the widespread adoption of spread spectrum will significantly improve most performance features of wireless communication in the future.

7 *Private ownership* of radio systems has overtaken all commercial, military and government use.

8 *Wireless technology* has been demystified in the minds of the users. For them all complexity has been removed, and they can just buy off the shelf, plug and play.

9 *Spectral occupancy* has moved from a perception of fully used and congested to grossly under-utilised and effectively infinite.

10 *Smaller wireless cells* have become increasingly economic and available to Joe Public at a very modest cost and almost free of set-up complexity and operation.

11 *Fixed–mobile network convergence* now seems to be on the horizon as a distinct possibility for the future.

12 *New forms of network* – permanent fixed and mobile, semi-permanent, sporadic and parasitic – have been made possible through points 1, 2, 6 and 7.

13 *New wireless real estate* can be made available and licence-free above 30 GHz due to the availability of new low-cost, high-performance devices, control and modulation schemes.

14 *More wireless means more fibre* and is necessary to provide the bandwidth to the places people and devices congregate. Without smaller cells of increasing bandwidth we stand no chance of solving the growing congestion problems.

15 *Direct wireless (radio) – optical conversion* is both possible and most likely a necessity in the future to continue the trend of wireless everything, everywhere, at continually falling cost and increasing utility.

## 9.4  What do People Want and Need?

It has been axiomatic for some time that in general people want and need to communicate at will from any location, and at any time, by a variety of means at a reasonable cost. That is – seamless access on the move! The dream is that we will be able to travel the planet and be able to connect to a global network infrastructure so we can work and play on demand.

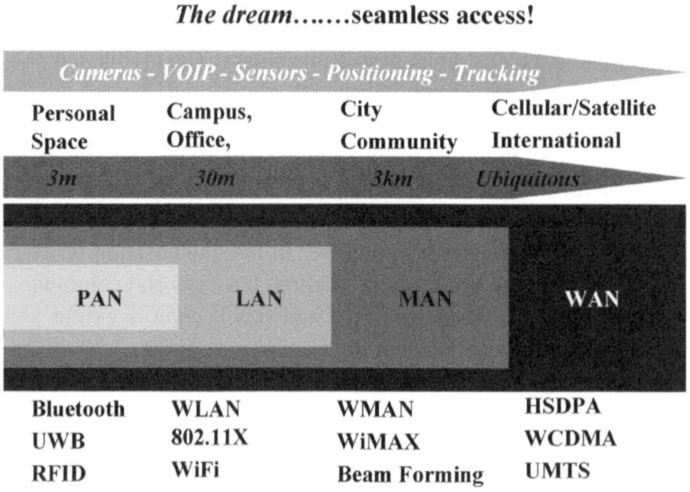

*The dream.......seamless access!*

| Personal Space | Campus, Office, | City Community | Cellular/Satellite International |
|---|---|---|---|
| *3m* | *30m* | *3km* | *Ubiquitous* |

| PAN | LAN | MAN | WAN |
|---|---|---|---|

| Bluetooth | WLAN | WMAN | HSDPA |
|---|---|---|---|
| UWB | 802.11X | WiMAX | WCDMA |
| RFID | WiFi | Beam Forming | UMTS |

*Cameras - VOIP - Sensors - Positioning - Tracking*

**Figure 9.4**   Seamless communication.

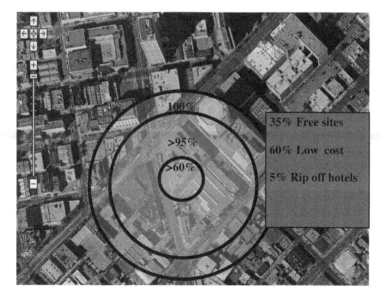

**Figure 9.5**   City block WiFi access.

To a large extent the mobile infrastructure of 2G, 2.5G and 3G has satisfied this need – you can indeed roam anywhere and get connected using GSM. But, all these systems lack the fundamental bandwidth required for twenty-first century work and play. Today it is WiFi that is the only available solution that fills the gap (see Figure 9.4).

In my personal experience I am always able to find WiFi access in any town and city within four blocks in any direction. But this is a relatively recent occurrence revealed by the lack of laptop connectivity to the net by a dial-up phone line for the past 2.5 years, and similarly for a mobile phone connection over the past 1.5 years (see Figure 9.5).

## 9.5 What can People Expect/Have?

There is no free lunch – someone always has to pay! But the good news is that there is no end in sight for the advance in technology and its availability at prices that will continually fall. So we can reasonably expect to have it all as depicted earlier in Figure 9.3. The key developments are most likely going to be in the areas of business models that provide a means of covering the cost of provision, operation and ownership.

The reality is that 'bit shifting' is already a commodity game. The fixed and mobile operators look more like supermarkets every day. Their profit margins get slimmer as equipment costs fall, staffing costs increase, and the customer willingness to pay fades. Further, competition intensifies as more players arrive. This will only increase in future as they all transition from Telco to Tesco – a 100% commodity game! Like food and clothing, connectivity and telecom services will be essentials of life that we need but take for granted, and we will be unwilling to pay for them.

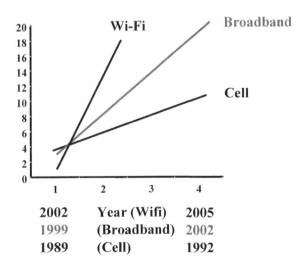

**Figure 9.6** Deployment rate and cost of mobility.

So the providers are going to have to look to the provision of new services, company consolidation, and the bundling of offerings to stay on top and remain profitable. I'm afraid the genie is well and truly out of the bottle in respect to the general public and companies understanding that bits and connectivity are relatively worthless (see Figure 9.6).

We can thus expect a number of big commercial developments in the mobile/wireless future:

- Massive consolidation exercises with ($\sim$)15 fixed and mobile providers reduced down to ($\sim$) three across most countries and continents.
- The coalescing of fixed and mobile operations as more mobile base stations means more fibre feeds.
- An acceleration towards infrastructure and base station sharing to reduce the capital and operating costs per operator.
- Integrated terminals providing 2.5G, 3G, 4G, WiFi, WiMax and BlueTooth from a single low-cost box.
- A huge increase in the number of services provided by the operating companies and those at the network periphery.
- Bundling of fixed, mobile, domestic, business, entertainment networks and services in and out of state/county/country.
- WiFi, WiMax and VoIP becoming a part of the main stream.
- A wholesale move away from circuit switching to IP-based networks.
- A massive reduction in billing complexity with 'all you can eat' fixed-fee provision dominating.
- More responsibility and initiative placed with the user rather than the major network operators.

## 9.6 Likely Technology Developments

Other chapters of this book deal with the general development of wireless technology, and so it is my purpose to look at some of the 'stage-left' surprises that will change the whole wireless scene that most will not have anticipated.

### 9.6.1 Home and Office

Let us start with a question; how many wireless transmitters do you own? For my household the list includes: five mobile phones, eight cordless phones, six walkie-talkies, five laptops, one PDA, three WiFi base stations, burglar alarm (= two mobile controllers = one fixed controller = 10 sensors = two alarm units), ten entertainment device controllers, four mice, four keyboards, two garage door openers, four sets of electronic car keys, four weather centre sensors, two VGA projector controllers, one laptop controller, one pair of headphones. And then there are the RFID tags in clothing and shoes! Frankly I have no idea what the actual number of wireless transmitter devices actually is. But it is over 75 for sure!

Over 75 wireless transmitting devices, fixed and mobile, for a home of four people came as a complete shock. Moreover, but for the death of my mother and departure of two of my children this figure could well have exceeded 100! Where did they all come from, and will I be seeing even more in the future? Yes has to be the answer.

In the same manner as electric motors in the home and car went from one in the early part of the last century, the microprocessor did the same at the latter part of the previous century, and it looks as though wireless devices are on the same trajectory. All of these technologies started life as big, bulky and expensive items that were soon reduced to commodities. Modern homes and cars typically have 30 to 100 (or more) motors and microprocessors, and soon this will be equalled or eclipsed by wireless devices.

So beyond my list above, where does the next raft of devices come from? How about entertainment? No woman I know likes to see wires, and entertainment systems mean lots of wires.

At least one of the major entertainment box makers will soon be introducing radios, TVs, DVD players, screens, projectors, hifi, games consoles, servers etc. all networked over 60 GHz (or thereabouts) wireless links. No wires! Just buy it, plug into the power, and all the boxes will talk to each other. Surround-sound speakers will connect to the amplifier, to the DVD player and games machine to the TV and/or projector *et al.*

The same brand PCs, laptops, PDAs, MP3 players, cameras *et al.* will follow to create an integrated appliance suite all connected wirelessly (see Figure 9.7).

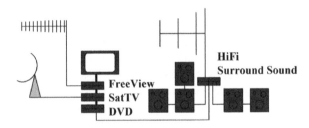

**Entertainment and IT disconnected, but configured using hard wired solutions. Nothing inter-works or talks together with ease, and there are wires everywhere!**

**Figure 9.7** The wireless home and office.

And then what happens? How about a link into the broadband network via DSL, fibre or WiMax, and the provision of further access/links WiFi + BlueTooth? Then for good measure, say they include an integral 2.5/3G base station with a 20-metre range so the owners can bypass their carrier's mobile network, and the picture is all but complete.

This is going to be quite a revolution. To say the least, the availability of low-cost 60-GHz TX/RX units will most likely open up a new lease of life for WiFi and will most probably change the operating model for everyone. Users, network providers, box suppliers and service suppliers will see a new, and in some ways even more converged world.

## 9.6.2 Manufacturing, Retail and Logistics

This area stands ready to be revolutionised by RFID. About 47% of the cost of everything we buy is down to transaction costs due to the inefficiencies in the processes of production, shipping and purchasing. To be blunt, the whole supply chain is a paper-heavy Dickensian process. RFID will change everything and reduce overall transaction costs dramatically.

From the mining, culling, cropping and gathering of raw materials, through transport and supply to processing plant, production units, wholesale and retail, customer purchase and aftercare, RFID tags will be able to identify and chart everything. From collection to customer, the source, processing and supply record will be on tap.

Let us consider just one facet of this technology at retail level. Apart from the obvious savings in logistics, accounting, shelf-fill and customer care, we might expect the electronic point-of-sale (EPOS) to be transformed. No more cash registers and perhaps no more people. Just walk through with a basket of goods and the RFID system totals the shopping and adds the amount to your mobile phone banking facility via BlueTooth.

Should this happen or not, RFID readers will be needed at every EPOS, and so will a broadband connection to deal with all the transactions. Now the chip-set at the EPOS will be identical to a PC or laptop and will come WiFi and BlueTooth ready. So why not provide high-speed net access to all the mobiles within range as they pass through. In fact, if the supermarket is suitably located, might they not also provide access via WiMax? So a new model might be: *buy over $100 of groceries from my store every week and I will give you free/reduced price net access.*

## 9.6.3 Logistics of Things and People

At a modest estimate, over $2000bn a year is lost by the inability of the container industry to efficiently manage the movement of goods across the seas and oceans. This number is most likely exceeded by the land portion of the same journey, and far exceeded by people movements. So what can be done?

RFID on all containers, boxes, packages and goods being moved is the first step in the solution. GPS on all vehicles is the next vital component so that everything can be located, identified and certified remotely and well ahead of need. Today the location of a seagoing container, box or package is almost always uncertain. Tomorrow it will available in real time. The whole logistic chain will be available for viewing and control from end to end. This will automatically lead to the better use of resources on land and at sea.

There are also security implications here regarding the future right of entry to a port for container ships. Specifications and laws are already being drafted in respect of minimal information standards concerning the origin and history of any container/load requesting entry into the USA for example. At a very minimum, shippers will have to be able to show that the load is legitimate, from a reputable source, and that there has been no interference en route.

All of this will introduce a new localised and significant traffic capacity, loading the wireless spectrum in and around ports, ships and the shipping lanes. And when extended to road carriers, the same will be true on land. We might then envisage gas stations becoming high-speed hotspots for transport data of a new nature over and above traffic information and WiFi access to the net.

### 9.6.4 Parasitic Networks

We now have to imagine a world where everyone and everything has a wireless device. There will still be locations where connectivity and bandwidth will still be a bit thin – such as in deserts, up mountains, in forests, at sea, on lakes and rivers, in canyons, underground parking lots, tunnels etc. There are always going to be locations that are signal-free, but people and things will still want to communicate. This leads to the notion of *parasitic nets* that configure spontaneously to provide a signal path via *daisy-chained* fixed and mobile people and things until a viable path is found.

The potential of this idea is not hard to imagine. At Dulles Airport recently I could not get a mobile signal, but a lady 20 m away near the widow was using her phone with no problem. So why couldn't I use her mobile as a repeater station? Why couldn't the network of mobile phones walking around the building conspire to give 100% coverage? At a recent conference there was only one CAT5 LAN connection and 20 or so people wanting to use it. So I plugged in, got online, and then turned on WiFi to make my laptop a local hotspot. Why not?

So here I am walking in the outback and wanting to send a text message. However, there is no cell signal! But soon several individual walkers and groups pass by going in different directions. So why can't I give them all a copy of my vital message so they can pass it on as soon as they get in range of a cell site, home WiFi or passing vehicle? I'm sure you get the general idea, and it is depicted further in Figure 9.8.

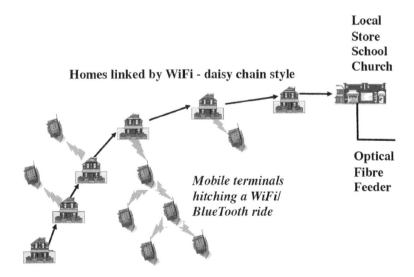

**Figure 9.8**  Parasitic networks.

Today we see people physically clustering to share movies, sound file and pictures on laptops, PDAs and mobile phones, at close range via BlueTooth. It is a very social and simple thing to do. Everyone benefits and no one worries about the cost of charging their battery or the use of memory.

A lot of hotels, companies and individuals provide free WiFi access for residents and visitors because it is just easier to do something positive than negative, it brings more business and makes life easier for everyone. Moreover, the cost is insignificant at around $1/day. Hotels don't charge separately for soap, shampoo and towels, which cost around $ 7!

Parasitic networks will work for the same reason. They cost nothing and benefit everyone. The mutual good outweighs the individual gain.

### 9.6.5 Mobile Sensor Networks

In 2007, all new mobile phones in Japan will have to have a GPS capability by law. So the network will know with great precision, for the first time, where all mobiles are located. The next logical step is the inclusion of a compass and accelerometer so the phone can be used as a pointing device.

By virtue of a series of inclusive nano-chips, a mobile will know where it is and where it is pointing. So it will be possible to say 'What is that building over there?' and then point your mobile phone and quickly extend you arm so the device knows position, direction and range (within limits). The network can then provide the answer – the White House!

This then is the birth of a raft of location-based services that people are likely to really value because of their utility in everyday work, play and travel. But consider further the range of sensors that could be included beyond this elementary stage. How about temperature, humidity, pressure, light intensity, background noise levels, carbon monoxide, radiation levels, toxic materials of all kinds etc., and suddenly we have the ultimate sensor network. Moreover, we might also like to consider a series of external (or integrated sensor devices) related to the long-term health and care of the individual – blood pressure, respiration, glucose level etc. In addition, we will be able to map their normal patterns of movement, travel and other activities.

When everyone and everything has a sensor capability we will, in principle, be able to see and qualify the air cleanliness and environmental friendliness at the very locations where people and things are present. We will also be able to monitor their activities and correlate to identify any significant changes. So the potential implications for transport, healthcare and security are profound.

## 9.7 Clusters of People and Things

When terminals were static and people's habits more random than socially correlated, the design of networks was statistically straightforward. But when the terminals started to move with the people a new and far more difficult causal environment was created.

For example, every day of the week coffee brings down mobile networks. Imagine a large conference – a gathering of over 500 people, say. They are all listening to the last speaker of the morning and (by and large) all mobile phones are switched off. At 10.30 coffee arrives and 300 mobile phones are switched on and the cell can't cope.

And it is not just coffee. Accidents on freeways, cancellation of flights and trains, phone-in radio and TV programmes etc. are all strange attractors invoking network congestion and call failure. It gets much more sophisticated as the terminals go beyond text and voice. When you get a meeting with a reasonable number of laptops on line at the same time the peaks of bandwidth demand can be huge. Correlated demand may be promoted by a few words in a presentation, the need to get the latest data, or someone offering their material for immediate download. The list of strange attractors is continually growing and may be endless.

In the old static telephone network the peak-to-mean traffic-loading ratio was around 4:1. By the way, this changed dramatically with the arrival with dial-up modems for web access. In the mobile network the same ratio is of the order 50:1, whilst the Internet sees numbers in excess of 1000:1 (see Figure 9.9).

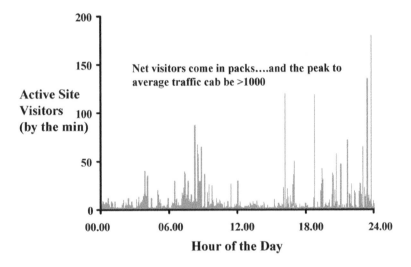

**Figure 9.9**   Network chaos.

There is no perfect solution, but creating micro- and pico-cells and hotspots will go a long way to a cure. When people and things cluster we need to provide huge amounts of bandwidth, and the technologies outlined in this chapter will do this in large measure. Beyond that we need to invoke protocols that differentiate between real time and non-real time, urgent and non-urgent.

The good news is that we have the basic tools, techniques and technologies to tackle this rapidly growing problem. Moreover, we have the basic engineering abilities to hand. But best of all, the natural line of developments are in this general direction anyway. We may not have to push water uphill!

## 9.8 Finally

I came from a world of copper wires and hot valves, with huge powers and a lot of heat. I conjured a vision of the world's network as a near infinite ball of string, with links between houses, villages, towns and cities spanning continents, seas and oceans. This extended to wireless with energy spread out or beamed into the ionosphere, across the surface of the Earth in a line-of-sight fashion, or beamed up to satellites to feed some distant terminal/s.

Today I meet young people who have never connected a computer or printer to a network with a piece of wire and a connector. All their lives have been in a wireless world. They have WiFi in the home and school, and some even have it in the car to tap into the 3G network. And so I try to see the world through their eyes. How do they envision networks?

So I have come to the conclusion that they, and we, should be thinking of a cloud of energy when we think of networks and networking in the future (see Figure 9.10). And as with any cloud system, the energy contained is not homogenous. It is lumpy, clustered, thin and thick, but mostly always there in some form.

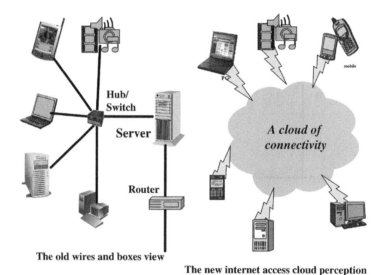

**Figure 9.10**   Wires or clouds?

The wireless future will therefore be a most unusual, but beneficial cloud, connecting everything and everyone, on the move or not. Ubiquitous communication will then have become a reality.

On what timescale is all this going to come about? In my view it will occur faster than perhaps we think, but some changes will seem glacial because of the sheer infrastructure investment required. In Figure 9.11, I have charted by best estimates and guesses for the time line of change. You will see that all the edges are hard – with things starting and stopping abruptly. In reality change rarely does this, it tends to build up slowly, accelerate, peak, and then hold for a while before decline sets in. Recent IT examples include floppy discs, VHS tapes, and soon it will be the CD and DVD that will be overtaken by a higher density storage media.

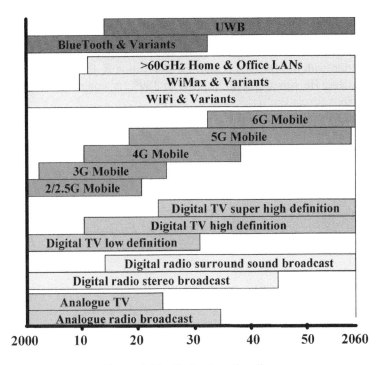

**Figure 9.11**   Technology time line.

So what I have tried to do is plot the steady state, dominant time periods of each technology on a global basis as I see it. To do this I have consulted numerous web pages, journals and papers, but at the end of the day we are all using a similar crystal ball. The only thing we can be sure about is that it will all happen. Why am I so sure? I have seen all the technology already in laboratories worldwide. And to be sure, there is even more to come!

## Biography

*Peter Cochrane, OBE, BSc, MSc, PhD, DSc, GGIA, FREng, FRSA, FIEE, FIEEE*
A seasoned professional with over 40 years of hands-on technology and operational experience, Peter has been involved in the creation and deployment of new technologies, the transformation of corporations, and the starting of many new businesses. Peter has done everything from digging holes in the road to working as a technician, engineer, educator, manager, innovator, technology prophet, and business angel.

His career in BT saw him progress to Head of Research and CTO with a 1000-strong team engaged in studies spanning optical fibre, fixed and mobile networks, terminals and interfaces, artificial life and healthcare, through to war gaming and business modelling.

Peter has also spent time in academia as an educator. He was appointed as the UK's first Professor for the Public Understanding of Science & Technology at Bristol University, UK, in 1998. A graduate of Nottingham Trent and Essex universities, Peter has received notable recognition with the Queen's Award for Innovation & Export in 1990 and several honorary doctorates. He was awarded an OBE in 1999 for contributions to international communications.

Peter currently serves on the board of companies in the UK and USA and his home page (www.cochrane.org.uk) is one of the busiest on the web.

# 10

# The Role of Ad-hoc Technology in the Broadband Wireless Networks of the Future

Gary Grube and Hamid Ahmadi

## 10.1 Introduction

Ad-hoc networking is a new paradigm that is already changing the way many business enterprises and public agencies use wireless networks.

By enabling network nodes to connect with each other in a peer-to-peer fashion – either to form a completely autonomous network or to extend traditional fixed infrastructure – ad-hoc networks open the door to new business models that lower costs while bringing wireless broadband coverage to areas that have previously been too difficult or too expensive to reach. Ad-hoc technologies also enable mobile nodes to self-configure to available networks, and for the networks themselves to automatically heal broken routes. Through these self-forming, self-managing capabilities ad-hoc networks promise to bring unprecedented flexibility and simplicity to mobile enterprises, as well as to traditional businesses and carriers that need an easy way to grow and evolve their wireless networks.

The advent and growth of the Internet profoundly changed the way people communicate, compute and do business – decentralising the control of information and empowering carriers, content providers and end users alike to create new opportunities based on their own visions of success. In the same way, ad-hoc technologies promise to enable a new generation of alternative wireless networking – decentralising the network through peer-to-peer connectivity, and giving carriers, devices and users new choices and opportunities in the fourth generation of wireless.

*Wireless Communications: The Future*   William Webb
© 2007 John Wiley & Sons, Ltd

## 10.2 The Need for Flexible Wireless Broadband Solutions

In the next 5–10 years, wireless broadband access will increasingly be regarded as an essential part of society's infrastructure, not just a tool for top-level business managers or a plaything for technology buffs. Consumers and rank-and-file businesspeople alike will increasingly depend on wireless broadband for a limitless variety of applications, from everyday tasks like finding a good restaurant to mission-critical jobs like securing a disaster scene.

And all along the usage spectrum – from personal leisure, to business connectivity, to mission-critical public safety and military applications – users will increasingly depend on seamless mobility. We're already seeing this trend today, as business travellers require convenient access to corporate data in airports, coffee shops, hotels and even across entire metro areas. In the same way that the Internet's value is fuelled by its own continuous growth, cells of wireless broadband coverage will continue to grow, overlap and integrate in order to support constant connectivity and information access – anywhere, anytime, on any device.

Looking just at the middle of the usage spectrum – the professional mobile workforce – IDC predicts a 20% growth in mobility over the next few years, totalling 878 million mobile workers worldwide by 2009. In the USA alone, the mobile worker population will reach 113 million [1].

The danger here is that a proliferation of wireless networks, relying on various standards and technologies, will drive up costs and complexity for carriers, enterprise IT and end users as discussed in Chapter 4. Using today's traditional wireless network models, the hazards and challenges associated with provisioning, supporting and using wireless broadband access can be expected to increase exponentially.

That's why the networks of the future must minimise the expense and complexity of extending coverage to new areas and providing new services that span across different networks. Broadband wireless networks, and devices that connect to them, will increasingly be required to provide a seamless experience.

### Anytime, anywhere, any service connectivity

Today, users must struggle with a bewildering array of options. Which network provides the geographical coverage they need? Which carrier provides the services they want to use, at the best price? What happens when they roam out of the service area? If more than one carrier is required to provide the needed coverage, how are connection handoffs and billing handled?

The ultimate goal should be the ability for users to connect any device with any service, anywhere and anytime, without worrying about any of these questions beyond the initial choice of a carrier or wireless Internet service provider (WISP).

To achieve this ubiquitous wireless connectivity, carriers and WISPs need to work together to achieve compatibility in both their networks and their supporting business practices. At the same time, device manufacturers need to design next-generation devices around industry standards and intelligent technologies that allow devices to recognise one another and interact flexibly in any context.

### Location-independent information delivery

The way Internet-based remote access works today, users have to know where to find the information they need – keeping track of connection types, logins, directories, URLs, inboxes, archives, personal folders and more. Add wireless access via innumerable networks around the world – with varying ownership, network types and interfaces – and finding remote information becomes even more of a chore.

Wireless broadband networks and devices of the future will need the built-in intelligence to keep track of the user's location, automatically discover and connect to the best available network, understand the user's information preferences, know where the information is located and when it has been updated, and deliver it to the user's device of choice seamlessly.

### A focus on the user experience, rather than the details of technology

If these connectivity and usability issues are complex today, the future will only bring more complexity through an explosion in the number and types of carriers, as well as in the variety of applications, services and options they provide.

Today's end users already have to deal with all the complexity they can handle. Successful wireless broadband carriers and WISPs of the future will be the ones who provide all the coverage, interoperability and features their users demand, while removing the burden of managing all these choices from IT departments and end users.

Moreover, successful designers of wireless devices, operating systems and applications will be the ones who incorporate autonomic control based on the user's explicit preferences as well as automated awareness of the constantly changing geographical and technical context. The goal is to hide the complexity of technology from the end user, enabling users to focus on the overall experience of seamless mobility and the specific tasks they want to accomplish.

### Persistent and secure connectivity

In creating a seamless, user-focused experience, one of the biggest challenges is to enable persistent, secure connectivity whether users are at home, at work or on the road anywhere around the world. Today's Wi-Fi networks provide persistent connectivity as users move from place to place – or from access point to access point – within the same network. But tomorrow's wireless broadband sessions must also persist as users move from network to network, and even from device to device.

For example, as you begin a typical day in the future, you might scan the latest news on your laptop, with stories automatically gathered and sorted to match your specific interests. When it's time to go to work, the news you were reading transfers automatically from your laptop to the car's sound system, and a voice begins reading where you left off. En route, the voice might alert you to an incoming call, which you take over the same sound system. Upon reaching your office, the call automatically transfers to your cell phone as you park and walk to the office. In the building, the call automatically switches from your cell carrier to the company's VoIP system.

Wireless broadband networks of the future will enable this persistent connectivity across business and pleasure activities, as well as all of the networks users encounter and the devices they use throughout the day. And as users come to expect persistent, ubiquitous connectivity, they'll also need the assurance that their devices, communications, corporate intellectual property and personal information are all secure, on each network encountered and across each connection handoff. This will again require cooperation on the part of carriers, providers and manufacturers to ensure consistent handling of security protocols and user preferences.

### Simple network extension and integration

As important as simplicity is for end users, it's just as critical for wireless carriers and WISPs to have a simple way to create, extend and integrate networks. Today's wireless networks are built on a 'star' topography, with one or more central hubs that route all messages to and from client systems. This architecture is extremely infrastructure-intensive, requiring provision of wireline or wireless backhaul to each access point.

But this model will become increasingly cost-prohibitive in the coming years. Most people living today have never even made a phone call, let alone logged on to the Internet. And even in urbanised areas of the industrialised world, competition and differentiation will lead to far greater density of coverage compared to today's relatively spotty wireless broadband access. At the same time, the user base will grow exponentially, while richer content and applications will consume ever more bandwidth.

To accommodate the explosive growth in demand for wireless broadband, carriers and WISPs need a much more cost-effective architecture for creating new wireless networks, as well as extending existing networks into new areas of wireless coverage. Through peer-to-peer connectivity, ad-hoc technology enables networks to form and grow in a much more organic fashion than the traditional star topology. Backhaul can be minimized as devices themselves act as intelligent routers to extend broadband connections wherever they need to be.

This could mean something as simple as extending a company's Wi-Fi network to a difficult-to-reach area without adding a new access point. Or it could mean something as profound as extending coverage to remote villages to offer people advantages they've never had before such as education, telemedicine, news and instant contact with distant friends and relatives.

## 10.3 Current and Emerging Models of Peer-to-Peer Broadband Connectivity

The evolution towards an experience of seamless connectivity and mobility is being driven by progress in wireless communications standards, radio technology, application interoperability and – most importantly – the architecture of the network itself. Although much work remains to be done, innovations in each of these areas are already enabling new business models and beginning to transform what's possible for consumers, government agencies, public safety workers and enterprises. Let's take a look at several representative examples.

### 10.3.1 Wireless Home Networks

Wireless home networks are beginning to offer peer-to-peer routing to easily move broadband content around the house without wires – and without requiring any individual device to be within close proximity to a content server or wireless gateway. In future years, we'll also see increasing use of pure peer-to-peer applications, with consumers sharing images and content directly rather than relying on a centralised content provider.

These peer-to-peer applications will be appropriate in homes with multiple simultaneous users, and will be especially well suited to public spaces, such as sporting and entertainment venues, where large numbers of people with similar interests congregate. For example, today's 'fantasy league' sporting games may shed the 'fantasy' aspect in the future, as spectators at live sporting events compete with each other and with the whole crowd, using cell phones and other mobile devices to predict plays and outcomes.

## 10.3.2 Military Applications

The US military was one of the first adopters of peer-to-peer technology, implemented as 'mesh networks' that can be established almost instantly in the field without the need to predeploy large towers, antennas or other traditional backhaul connections. Mesh networks provide troops with instant broadband communications across the battlefield – improving surveillance, tactical planning, targeting accuracy and troop safety as shown in Figure 10.1.

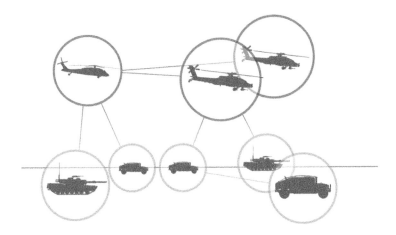

**Figure 10.1** The military was one of the first to commission robust, self-healing ad-hoc wireless broadband networks to allow high-speed, high-bandwidth communications for maximum survivability in this harsh environment.

Every soldier's radio powers the network and helps route signals efficiently, with minimal loss to signal strength. Peer-to-peer technology in a mesh network creates an interconnected web of radios that automatically extends network coverage and robustness as new users join the mesh. In addition to basic voice and data, military mesh networks also support real-time data and video connectivity, giving field commanders unprecedented insight into situations as they develop.

## 10.3.3 Public Safety

With the success of mesh networks in military applications, public safety workers – such as police, fire-fighters, search and rescue, hazmat and other first responders – are quickly adopting similar technology. Mesh technology allows public safety workers to establish and extend networks at the site of an incident simply by bringing mesh-enabled devices to the scene.

In public safety applications, ad-hoc technology allows delivery of broadband data between an incident and the command centre without traditional backhaul connections to each access point, and without dedicated routers/repeaters. Using mesh networks in this way, public safety agencies can significantly lower operating costs while enabling officers to download maps and mug shots, send live video to dispatchers, monitor live video from security cameras, and more.

Mesh networks also enable peer-to-peer relationships to form automatically between handsets and other personal devices, so first responders can instantly communicate and share data on the scene of an incident. Through dynamic routing of signals between devices, and boosting of signal strength at each device, mesh networks can overcome distance and line-of-sight limitations that often hamper traditional wireless solutions in difficult environments.

For example, fire-fighters can carry small peer-to-peer communication devices on their person – and if necessary drop temporary wireless access points and repeaters in strategic areas – in order to stream images and information to one another and to the incident command centre. Peer-to-peer technology can even incorporate triangulation and timing capabilities to provide relatively precise location tracking in areas where GPS cannot reach, improving safety for first responders as they work to secure the safety of citizens.

## 10.3.4 Private and Public Transportation

Departments of transportation are building intelligent transportation systems that improve traffic safety and reduce congestion by enabling smart signals and monitors to communicate directly with each other to optimise timing, warn of danger, and communicate information to the command centre for strategic planning. Ad-hoc technology enables rapid deployment of these systems without the disruption and expense of laying fibre-optic cable to each signal and sensor.

In the future, the same technology will be used to extend these transportation networks into vehicles on the road, directly enhancing the safety and convenience of individual drivers. One manufacturer already offers cars that can 'talk' to other vehicles of the same make to warn of upcoming hazards, and other car makers will undoubtedly follow suit. In the future, cars will automatically suggest the safest and fastest routes, sense one another as well as guard rails and other highway features, and even take corrective action automatically in the event of a sudden emergency. Should a crash occur, police and medics will be notified automatically of the location, number of cars involved, impact severity and other factors.

And in normal driving, the same networks will provide the bandwidth and connectivity for directional information, safe forms of entertainment and communication, and other uses. Because ad-hoc technology supports automatic connectivity, connection handoffs and multi-hopping of signals at vehicle speeds over 200 mph, it will be much easier and more cost-effective to extend today's intelligent transportation systems using this flexible architecture – rather than covering extensive highway systems through backhaul-intensive fixed architecture.

The same high-speed, persistent mobile connectivity that interests transport departments and car manufacturers will also enable rail, subway and bus transit agencies to improve operational control and safety. Ad-hoc technology can be used to extend today's sporadic or even nonexistent broadband coverage to span entire transportation systems cost-effectively. And the same constant connectivity that gives drivers and traffic controllers better tools to improve efficiency and safety can also be used to provide riders with communication, information and entertainment – making their commutes far more enjoyable and productive.

## 10.3.5 Metro-area Broadband Networks

Many local governments and private enterprises today are deploying multiple Wi-Fi hotspots in an attempt to provide broadband wireless coverage across the core business and commerce areas of major cities. Increasingly, however, the goal will be to reach beyond the distance limitations of traditional Wi-Fi to provide true metro-area broadband coverage across entire urban areas.

While there are various architectures to support metro-area wireless broadband, ad-hoc technology provides clear advantages in terms of cost, flexibility, reliability and scalability. Peer-to-peer connectivity can be used to extend the reach of broadband wireless while minimising new investments in backhaul infrastructure. With ad-hoc technology, it becomes cost-effective to cover residential urban and even suburban areas – instead of covering only high-value business and commercial districts.

Unlike traditional wireless architectures, ad-hoc architecture also provides the resilience to enable wireless communications even when cell towers and other infrastructure have become inoperable due to storms, sabotage or malfunction. Police, fire departments, transportation departments and other public agencies – which today typically rely exclusively on their own private networks – will also be able in the future to 'piggyback' on these metro-owned wireless networks. This will provide backhaul for their own peer-to-peer networks, greatly extending their reach and capability.

## 10.3.6 Mining and Manufacturing

Mining companies and other enterprises that operate in extremely difficult work environments are looking for new ways to improve communications and safety. Stringing cable throughout a mine is impractical, expensive and potentially dangerous. Wireless signals from a central access point cannot penetrate rock walls. GPS location systems require an unobstructed view of the satellite. These issues and more arise from the relatively inflexible nature of traditional architectures, built on the concept of fixed nodes and line-of-sight wireless communications. Ad-hoc networks overcome all these limitations by allowing communication signals and location information to multi-hop from device to device, around obstacles. In a mine, a small device can act as both a communications and location system for each miner, travelling with miners wherever they go and ensuring continuous contact through sophisticated timing, triangulation and dynamic routing capabilities. Manufacturers can use these same capabilities to gain visibility and control over every aspect of their business – people, supply chains, tooling, distribution and even the performance and maintenance needs of finished products.

## 10.3.7 Corporate Networks

Enterprises can use ad-hoc technology to extend wireless connectivity to hard-to-reach areas of the corporate campus. Older buildings can be difficult to wire; corporate campuses may have satellite buildings that are difficult to reach with Wi-Fi; and some workgroups may be too small to justify the expense of adding an access point. For these reasons and more, extending the enterprise wireless network through access points built on the traditional star architecture may be costly and impractical, or even impossible.

Ad-hoc networks can provide whole-campus coverage for high-definition video, VoIP, Internet and business applications, using peer-to-peer technology that minimises the need for wired access points and repeaters. And with the emergence of cognitive radio technology as well as multi-mode capabilities – for example, incorporating both cellular and peer-to-peer technology in a single device – tomorrow's workers will increasingly have continuous, automatic access to the best available network. Without even thinking about finding a wireless network and configuring to it, users will have full voice and data access to stay productive wherever they go.

### 10.3.8 Sensor Networks and Things-to-Things Communication

Low-power sensors and smart devices are already enabling things to communicate directly and intelligently with other things. This is a new paradigm – an extension of the Internet into the world of things. When people think of the Internet, they generally imagine people using it to gather information and enjoy content that was ultimately generated by other people – bloggers, retailers, game programmers, musicians and so on. But ad-hoc technology and other alternative technologies are already bringing intelligent communication capabilities to things, and this emerging trend is poised to change the way we interact with the world and perform all kinds of task.

For example, one automobile manufacturer is using intelligent two-way sensors on vehicle components such as wiring harnesses, which become too brittle for installation when they are cold. The sensors report the temperature and location of automotive components, enabling supply chains to automatically adjust for optimum delivery times. Another example is industrial agriculture, which uses remote sensing and control to manage irrigation, fertilisation schedules, pest control and more. Ad-hoc technology can be incorporated in these sensors to create wide-area networks across the farm, providing greater visibility over the entire operation while saving labour costs through autonomous control.

Intelligent sensor networks will increasingly play a role in homeland security, traffic control, energy and utilities, media, manufacturing supply chains and countless other government and business fields. Beyond the enterprise, sensor networks and things-to-things communication will transform personal and family life as well – automating home systems, security and home monitoring, entertainment systems, travel and tourism, social networks and other aspects of life in ways we can scarcely imagine today.

## 10.4 Enabling the Next Generation of Ad-hoc Connectivity

As the preceding examples illustrate, early adopters are already using ad-hoc networking technology to transform specific facets of home life and enterprise efficiency. In the years to come, these isolated ad-hoc deployments will expand, overlap and interconnect to affect nearly everything we do at home, at work, on the town or on the road.

This transformation will depend on flexible, scalable, low-cost wireless broadband architectures that enable peer-to-peer connectivity and autonomous control. These architectures will be essential for achieving pervasive 4G (fourth-generation) networks that have the following functionality.

***Take the wireless world to the next level of capability and value***
Wireless networking is still in its infancy. Today, you might be amazed at how easy it is to download your email wirelessly at a local coffee shop – but consider how limited you really

are. You might lose your connection simply by going to the bakery next door. And you still have to log on to your email server and check it manually – there's no way for email to track down your location and notify you that it has arrived. Pervasive wireless connectivity, based on intelligent devices that can auto-configure and communicate directly with one another, will open up countless new business opportunities and change the way people live, work, socialise, travel, learn and play throughout the day.

### *Bridge the gap between existing wireless infrastructure and emerging wireless needs*
One of the main tasks in enabling 4G networks that offer pervasive, seamless wireless connectivity lies in extending today's wireless infrastructure to meet tomorrow's needs. While fixed infrastructure will always be part of the equation, ad-hoc technology provides a way to extend wireless connectivity to reach new areas of coverage. Expanding coverage by adding fixed access points and wired backhaul will become increasingly cost-prohibitive and impractical as users expect nomadic and truly mobile connectivity.

### *Enable rapid reconfiguration and redeployment of resources*
In the same way that ad-hoc technology enables rapid, cost-effective extension of wireless to new coverage areas, it also enables rapid reconfiguration and redeployment of resources to adapt to changing coverage and bandwidth needs. Because ad-hoc technology allows fluid, automatic rerouting of signals based on device location and environmental conditions – rather than investment in a completely fixed routing infrastructure – it's the ideal technology to support wireless networks that are in a constant state of change. Just as the military increasingly relies on mesh networks to keep mobile units connected, ad-hoc technology will support new levels of business and personal mobility in the future without the time, expense and commitment of traditional network build-out.

### *Provide consumers and professionals with new options for content sharing and access*
The centralised 'star' architecture of traditional wireless is optimised for a subscriber-based model of network access and content distribution. The carrier or WISP offers a menu of services, providing access and billing according to the consumer's preferences. If consumers don't like the mix of services offered by one provider, they can shop for another. While this will continue to be the dominant model, ad-hoc technology will add a decentralising and democratising element – enabling access across providers, or even allowing users to interact and share content directly with one another, in many cases eliminating the middleman entirely.

### *Bring full connectivity to the world*
It can be shocking for people in developed countries to learn that most people in the world today have never even made a phone call. Ad-hoc technology offers the most cost-effective way to expand wireless coverage to reach rural and undeveloped regions of the world. With the ability to talk, share information and engage in commerce with the rest of the world, these underserved areas can transform their local economies and join the global economy – generating new wealth, resolving political differences and improving the lives of ordinary citizens.

## 10.5  Types of Ad-hoc Network

There are currently two main types of wireless broadband ad-hoc network: autonomous peer-to-peer networks with no fixed infrastructure, and hybrid mesh networks that combine ad-hoc features with fixed infrastructure. The two types are compatible and will increasingly overlap and integrate as wireless networks are built out to create ever-wider areas of coverage. At the same time, each type has its own strengths for particular providers, users and applications.

### 10.5.1  Autonomous Peer-to-Peer Networks

Autonomous peer-to-peer networks are constructed entirely of cell phones, PDAs, laptops and other mobile and personal devices as shown in Figure 10.2. When these are designed with built-in ad-hoc technology, all of these devices can serve as intermediate routers to create communications paths via multiple 'hops' from one device to another. This 'multi-hopping' capability is based on the ability of devices to recognise one another, automatically join to form a network, and evaluate signal strength and other RF conditions to create the optimal data path at any given moment.

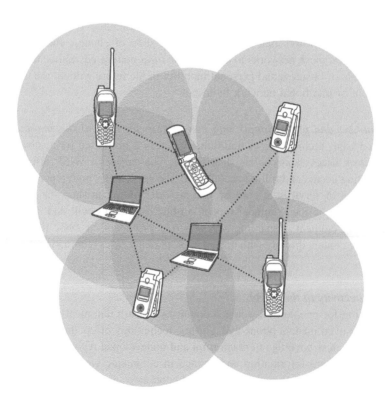

**Figure 10.2**   Illustration of an autonomous peer-to-peer network.

Autonomous peer-to-peer networks are ideal for the military, first responders and other highly mobile users who need the ability to create a fully functional wireless network anywhere and anytime, often in difficult field conditions, without relying on fixed infrastructure or time-consuming network configuration. Increasingly, the total flexibility offered by this type of network will become attractive to mainstream enterprises, enabling new business models based on unprecedented flexibility and mobility.

## 10.5.2 Hybrid Mesh Networks

Hybrid mesh networks include both ad-hoc-enabled mobile devices and fixed infrastructure elements such as intelligent access points and wireless routers, as shown in Figure 10.3. The mobile devices and fixed infrastructure cooperate to create the optimal route, using the same dynamic route discovery that autonomous peer-to-peer networks use. In addition to routing, the fixed infrastructure may also provide wired or wireless backhaul for connectivity to different ad-hoc networks, the enterprise LAN or the Internet.

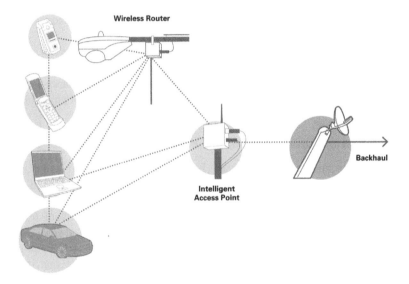

**Figure 10.3**  Illustration of hybrid mesh network.

Because one backhaul-enabled meshed access point can serve multiple local fixed or mobile wireless routers, which in turn serve local client devices, hybrid ad-hoc-enabled infrastructures can vastly reduce the number of fixed backhaul connections and significantly lower total costs.

This type of network can be ideal for providing wireless coverage across a corporate campus, coordinating sensors and signals in a smart transportation system, controlling manufacturing processes, extending the reach of traditional wireless networks and so on. Enterprises that don't need the ability to instantly set up a wireless network anytime and anywhere, or that require backhaul connections, will find this type of network to be a very attractive and flexible alternative compared to a build-out based on new fixed backhaul connections and access points.

## 10.6 Integrated Ad-hoc and Wide Area Networks

In addition to the two main types of ad-hoc networks – autonomous peer-to-peer and hybrid mesh – ad-hoc networks can also be integrated with a traditional wide area network (WAN). Properly implemented, this integration can improve the economics and performance of both the WAN and the ad-hoc extensions to it.

### 10.6.1 Linking of Ad-hoc Workgroups

In an autonomous peer-to-peer network, for example, workgroups must be in relative proximity in order to communicate. With enough devices spread over a wide enough area, there's no theoretical limit to the distance the network can span. But in practice, a workgroup in New York is unlikely to be able to communicate with one in Los Angeles using purely ad-hoc technology, because there won't be sufficient devices to multi-hop a signal from coast to coast.

However, the two groups can easily be linked together through a WAN, with ad-hoc technology extending the reach of the fixed network to mobile workgroups in each location. Any number of mobile workgroups can be linked together over any distance via the WAN, flexibly expanding the ad-hoc networking experience beyond the limits of physical proximity.

### 10.6.2 Extension of carrier broadband networks

Similarly, ad-hoc technology can be used by traditional carriers to supplement more infrastructure-intensive backhaul methods such as T1, fibre optics, 802.16 and proprietary microwave. While carriers will continue to use these traditional backhaul technologies, they will also increasingly rely on ad-hoc technology to provide backhaul for extending coverage to remote and difficult-to-reach areas. In this type of deployment, ad-hoc technology can offer carriers significant savings compared to traditional network build-out, while providing much greater routing flexibility to target new customers beyond the reach of established infrastructure. And by allowing rapid, cost-effective expansion of coverage, ad-hoc extensions also allow carriers to better serve their mobile customers.

### 10.6.3 Enhanced Network Performance

Beyond providing a practical and cost-effective way to link distant groups and extend coverage, integration of ad-hoc and WAN technology can significantly improve the operation and performance of each type of network.

When ad-hoc networks are integrated as extensions to a WAN, the WAN's fixed routes and dedicated routers can supplement the dynamic route discovery performed by the ad-hoc devices to improve the overall speed of route discovery. The WAN can also offer enhanced brokering for peer-to-peer transactions to support network security and user privacy, as well as to ensure proper digital rights management so that content creators are compensated fairly for their work. And the WAN greatly expands the pool of potential connections while facilitating the identification and organisation of interest groups to take advantage of these connections.

In turn, ad-hoc technology can enhance the performance and operation of the WAN by offloading a large percentage of high-bandwidth communications to the peer-to-peer relationships between devices. This frees up bandwidth on the WAN itself, while at the same time extending its coverage and increasing the mobility of end users.

## 10.7 Enabling Technologies

Up to this point, we've said a lot about what ad-hoc technologies are capable of doing and the benefits they provide. Now we take a brief look at what the core enabling technologies are and how they work.

### 10.7.1 Self-configuration and Self-organisation

Ad-hoc networks must be able to self-form, self-manage and self-heal with minimal human intervention. Mobile users cannot be expected to interrupt mission-critical work to configure their connections or troubleshoot problems, and IT staff cannot be expected to travel to every site where an ad-hoc network might be used.

Devices need the ability to recognise when an appropriate network is within range, and configure to it automatically. If a node goes down, other nodes must recognise the problem and instantly reroute signals around the broken link. Nodes must have the ability to handle communication and control functions on a peer-to-peer basis, with most human interaction being limited to exception-based commands. In short, ad-hoc networks must handle most technical issues autonomously, leaving users free to concentrate on the work at hand.

### 10.7.2 Multi-hopping and Dynamic Routing of Data Packets

One of the most important technologies for ad-hoc networks is the ability to automatically route data packets from device to device, dynamically adjusting the route to adjust for user movement, environmental conditions, available devices and other factors. This 'multi-hopping' technology is what enables ad-hoc networks to form, change and grow in the field without the need to provision additional fixed infrastructure.

The same multi-hopping capability can be used to extend traditional wireless networks with minimal new infrastructure investments. For example, mobile relays can be placed strategically throughout an urban area to boost signal strength and data rate as a signal is passed from a centralised access point to reach users on the outskirts of the city. The need to run backhaul to access points throughout the city is minimised, because these mobile relays can use their own in-band capacity to provide wireless backhaul. Carriers that in the past were forced to deploy microwave or wireline-based backhaul in order to expand coverage will now have a much more cost-effective way to reach new users and increase revenues.

### 10.7.3 Smart Sensors and Devices

We've already touched on the fact that tomorrow's wireless networks will incorporate things-to-things communications. The intelligent interaction of end user devices will migrate to smaller, increasingly autonomous sensors and devices.

For example, the home of tomorrow will have sensors that keep track of your body temperature, ambient temperature, and the status of your heating and ventilation systems and window blinds to automatically control everything for optimum comfort. The sprinkler systems of tomorrow will automatically sense soil moisture and temperature, adjusting watering schedules and patterns to balance water conservation and lawn health. The refrigerators and kitchen cabinets of tomorrow will keep track of all the ingredients on hand, suggesting recipes, warning of expiration dates and creating shopping lists.

Moving beyond the home, a visit to the doctor's office of tomorrow might even involve ingesting microscopic sensors that monitor organ function and scan for disease states. Tomorrow's highways might incorporate millions of sensors within the lane paint to detect when a vehicle is about to leave the roadway and automatically correct the steering. Tomorrow's national security might involve scattering millions of microscopic sensors along borders to track the movements of people and detect hazardous substances. With smart sensing and communications at the micro and even nano level, the possibilities are limitless.

In the future, smart sensors and devices will be deployed by the billions, perhaps even the trillions. IPv4, the packet-switching protocol that is most widely used today to provide a unique address for every Internet node, supports 4.3 billion IP addresses – less than one address for every person on the planet. Widespread adoption of IPv6, which potentially provides unique addresses for as many as $3.4 \times 10^{38}$ nodes – or more than one address for every atom in the Earth's continents and oceans – will be required to support the enormous growth in sensor networks in the years ahead.

### 10.7.4 Location-awareness

True mobility will be the hallmark of future wireless applications. Beyond the basic ability to stay connected while in motion, many mobile applications – such as emergency rescue, personal navigation, social networking, targeted marketing and more – will also benefit from the ability to know exactly where devices are located. Incorporating GPS in devices is a good start, but GPS doesn't work in tunnels, buildings and other places where there's no line-of-sight contact with GPS satellites.

Fire-fighters and other public safety personnel are already adopting devices that use precise timing information, triangulation and multi-hopping to provide device location and tracking capabilities in places GPS cannot reach. In the future, both GPS and timing/triangulation technologies will be incorporated into a variety of applications, from supply chain management to personal city tours delivered to your cell phone.

### 10.7.5 Low-power and Energy-scavenging Technologies

As the number of sensors and devices in deployment explodes, there has to be an alternative to sending IT people out to change all the batteries. Extremely energy-efficient designs will be required for many types of devices, while new energy-scavenging technologies will enable some types of nodes to operate indefinitely without batteries.

For example, an emerging generation of photovoltaic technology will allow so-called 'solar' cells to operate using the ambient light in your house, and the cells will be small enough to incorporate invisibly in windows and even paint. Other technologies will enable nodes to scavenge energy from thermal energy, kinetic motion, and other environmental sources.

## 10.7.6 End User Control over Preferences and Privacy

As wireless networks and applications proliferate, and as connections and services become automatic, end user acceptance will depend on providing a sense of complete control over preferences, affiliations, location information and other elements of personal security and privacy. These security and privacy controls must operate on two levels – enabling network infrastructure to be shared while preventing unauthorised users from controlling applications or accessing protected information.

In a corporate environment, for example, the fire detection and suppression system might share network bandwidth with the building-control systems, computer management systems, and even the copier monitoring and maintenance systems. You don't want the copier guy to set off the fire alarm. Nor do you want hackers to access personal information or protected content using tomorrow's ubiquitous wireless networks. To keep information segregated for authorised access, wireless applications will need to incorporate secure tunnelling. This is similar to the VPN technology used on today's Internet, but it will operate at the sensor level. Applications must give users the ability to specify their security, content and other preferences actively or by exception. And beyond these technical issues, it will be important to educate users that they can trust tomorrow's applications to protect their private information from being accessed without their knowledge and permission.

## 10.8 New Business and Usage Models

Of course, the whole point of all this new wireless technology is to enable new business and usage models – boosting the productivity of mobile workers, creating new revenue opportunities for business and improving consumer lives. Several of the most promising opportunities for emerging and future ad-hoc networking applications are noted below.

### Community Wi-Fi
Many local governments are already implementing public Wi-Fi access in city centres to provide Internet access at low or no cost to taxpayers. Community Wi-Fi can pay for itself by attracting advertisers and sponsors, as well as by encouraging businesses to locate and expand within the covered area. And by maximising this coverage area, municipalities can maximise the benefits to citizens and businesses alike. In the past, this has typically meant adding new Wi-Fi access points with dedicated backhaul to each one. But as ad-hoc technology becomes widespread it will become much more cost-effective to extend coverage using mobile relays as well as peer-to-peer multi-hopping in connected devices – significantly lowering total backhaul requirements, as shown in Figure 10.4.

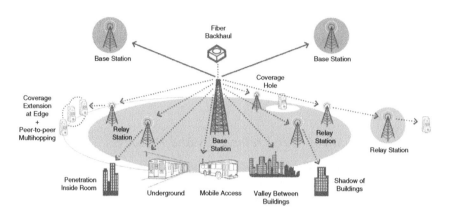

**Figure 10.4** Illustration of mobile, multi-hop relays that will enable greater coverage and range enhancement within cities, towns and other municipalities.


### Convergence of voice and data networks

Today, people generally have to keep track of what kind of network they're using, switching from cell phone to laptop as they use voice and cellular networks. But with the ability to build out high-bandwidth wireless networks affordably, carriers and WISPs will increasingly be able to offer a full package of wireless services – including voice and data – without requiring users to choose the right device and network, configure connections or reconnect when roaming between networks. Ultimately, the distinction between voice and data networks will become meaningless to end users, as they simply choose the device they prefer to communicate with and access all the information they need, without worrying about network types.

### Convergence of fixed and mobile usage models

By the same token, users will increasingly be able to get seamless access wherever they may be. Ad-hoc technology allows wireless broadband networks to grow, overlap and integrate – and encourages carriers and WISPs to forge business and technical relationships that allow users to roam freely without worrying about the underlying complexity. The ultimate goal is to provide one wireless umbrella of broadband coverage for fixed, nomadic and highly mobile users.

### Mobile TV and multimedia services

Persistent broadband connectivity will enable television and other rich content – which in the past has been limited to fixed devices – to seamlessly follow users as they move between different locations and devices.

### Data and voice services for rural communities and developing countries

Ad-hoc technology provides a light-infrastructure, low-cost way to deliver voice, Internet, remote education, telemedicine and other services to regions that have been left behind in the digital revolution. Extending connectivity to areas that don't even have basic phone service today will open up countless new business opportunities, while modernising economies and improving the lives of literally billions of people.

## *Public safety*

We have already discussed how police, fire-fighters and other first responders are using mesh networks to improve operational communications in the field. The extension and integration of wireless broadband networks in the future will expand the scope of these networks far beyond the scene of a particular incident, improving communications with the command centre and coordination between multiple sites. And as network density increases, so will available bandwidth – enabling delivery of live video and other broadband data without the need for extensive fixed infrastructure, time-consuming setup or expensive proprietary equipment.

## *Farm and industrial management*

Many farmers are already using remote sensing and control to manage irrigation. In the future, smart sensors will form wireless wide area networks across the entire farm, giving visibility and fine-tuned control over all operations. Water rates will be determined by actual soil moisture in each plot, weeds will be monitored and controlled with precisely the right amount of pesticides, fertilising and harvesting schedules will be determined by actual leaf colour and temperature, and so on. Manufacturing and other industrial operations will use sensor networks in a similar fashion to monitor and control processes from end to end.

## *Component and product tracking*

Companies are already using RFID tags at the pallet and even the product level to streamline the supply chain. In the future, smarter tracking capabilities will make products more useful and easier to maintain throughout their lifecycle. For example, if your plasma TV develops a problem, sensors embedded in the TV might automatically notify the manufacturer, track an authorized person as he enters your house to pick up the TV, and track the unit as it makes its way to the service centre and back to your home. Home, manufacturer, delivery company and service centre networks would all recognise the TV sensors and cooperate with one another to ensure prompt, safe service – with minimal hassle on your part.

## *New information and entertainment options*

In addition to streaming broadband content throughout the home and around the yard, the wireless broadband networks of the future will increasingly give consumers new interactive experiences in public places. For example, imagine a sporting event in which spectators can play fantasy games with one another by using their cell phones to predict the next play, and instantly see how their predictions match up with other spectators in the arena.

## *Social networking based on proximity and common interests*

Building on the last example, ad-hoc social networks will enable people to meet and interact with each other, sharing their common interests in new ways. For example, imagine you're visiting a favourite coffee shop, when your phone notifies you that a buddy and fellow football lover is passing by. You get together and check your phone for other football fans nearby. Soon you've soon assembled a lively crowd to discuss this weekend's big game.

## *Building blocks for the extended Internet (X-Internet)*

Increasingly, Internet and wireless technology will be used to enable direct, things-to-things connectivity and control. Initial opportunities for low-power, low-data-rate things-to-things connectivity will be in areas such as manufacturing and utilities, where labour can be saved and processes improved by automating the tracking of components on the factory floor, or the remote sensing of pipelines and substations. Implementations like these are already in deployment or emerging. In the near future, these models will be expanded greatly – for example, tracking products through the entire supply chain to speed delivery and lower cost. Eventually, the X-Internet will transform home comfort and entertainment systems, business processes, shopping, the social environment, security systems and more – virtually every aspect of life.

## 10.9 Benefits of Ad-hoc Technology Wireless Carriers and Internet Providers

The alternative wireless broadband technologies we've been discussing have the potential to touch virtually everyone on the planet, to the benefit of all. In terms of business profitability, the companies that deploy these new technologies to serve new users in new ways probably have the most to gain of all. We wind up our discussion with a look at how wireless carriers, cable broadband operators, WISPs, start-up wireless companies and interactive marketers can use next-generation wireless broadband technologies to create profitable new business models.

### 10.9.1 Incumbent Wireless Carriers

It's probably to be expected that incumbent carriers – who have spent many years and enormous budgets building out 2G and 3G cellular networks based on backhaul-intensive star architectures – might feel threatened by new, decentralising wireless technologies. But that fear is unfounded. As with any disruptive technology, the key is to learn how to harness it to extend or evolve the current business model to achieve an advantage.

Rather than making existing infrastructure obsolete, ad-hoc technology gives established wireless carriers a new opportunity to leverage their existing investments, extend their reach and offer profitable new services. Investing in backhaul to reach new coverage areas or to support greater mobility is an expensive proposition. However, ad-hoc technology gives existing wireless carriers a much less expensive, highly flexible way to provide in-band wireless backhaul to reach underserved and remote areas.

By adding ad-hoc technology to their existing networks, incumbent carriers actually have a head start on the competition. Ad-hoc technology gives carriers an affordable, fast-track pathway for extending their wireless reach to provide 3G and 4G solutions for converged networks, nomadic and mobile coverage, multimedia and a variety of value-added broadband services.

### 10.9.2 Cable Broadband Operators

The rise of cable TV offered a powerful alternative business model to broadcast media, charging subscriber fees for point-to-point delivery of a much greater selection of content without RF attenuation and interference. With the advent of desktop computing and the Internet, it was only natural for cable providers to leverage their existing wired infrastructure to offer broadband connectivity as well.

However, the existing cable model is completely unsuitable to meet the needs of users who are increasingly looking for ways to mobilise their entertainment and computing experiences.

To remain competitive, basic wireline carriers and multi-service aggregator/providers know that they'll need to continue serving fixed users while also reaching out to the growing mobile market.

Ad-hoc technology allows these enterprises to extend their networks to serve both fixed and mobile markets with a single broadband wireless technology. Ad-hoc solutions allow carriers to expand coverage without laying new cable – eliminating the 'last mile' problem. And ad-hoc solutions provide better economics than 3G wireless solutions, scaling up easily and gaining performance with greater user density. By adding ad-hoc technology, cable operators can cost-effectively extend their reach to outlying areas, and even more cost-effectively add bandwidth to respond to a rapidly growing subscriber base in urban areas.

### 10.9.3 'Mom and Pop' Wisps

Small wireless Internet providers in rural and underserved areas need an affordable, small-scale solution that incorporates low-cost, alternative base stations. They simply cannot afford the massive infrastructure investments that large, name-brand providers have already made, given that their investments must be recouped from a much smaller subscriber base.

Using ad-hoc technologies, these 'mom and pop' WISPs can enter a remote area and begin providing broadband services very quickly, with minimal investment in fixed infrastructure. Ad-hoc networks scale down very well compared to cellular and other wireless network types – especially for the fixed broadband that is likely to dominate in rural areas for the foreseeable future. And they can be configured to provide coverage over a relatively large yet well-defined area, without relying on wired infrastructure that may be spotty or nonexistent.

Ad-hoc networks are also ideal for small, alternative providers because they easily adapt to whatever spectrum is available in a given area. For example, in the USA, the 700 MHz band is not being used for TV signals in many areas, and will eventually be completely auctioned off for different purposes. Mom and pop WISPs can use this slice of spectrum to create 802.22 wireless regional area networks that bring broadband to areas not easily served by cable or DSL.

### 10.9.4 Greenfield Operators

In many areas, the market is ripe for start-up enterprises to leapfrog into the wireless business across multiple degrees of difficulty. These 'greenfield' operators (the term has its origin in agricultural fields that are considered ripe for expanding urban development) have no existing broadband infrastructure. Because they're not tied to any legacy technology, the advent of ad-hoc technologies gives start-up operators the quickest, easiest and most affordable way to build and expand a broadband wireless infrastructure to meet the needs of fixed, nomadic and mobile users. Solutions based on technologies like OFDM and protocols like WiMax enable greenfield operators to serve emerging wireless broadband markets directly and cost-effectively, minimising the need for backhaul.

### 10.9.5 Marketers

When content is distributed over an ad-hoc network, as opposed to traditional broadcasting, there is a communications path back from every device. The ability to gather information on this reverse channel – within the privacy parameters established by the user – enables marketers to create new business models based on targeted understanding of individual users and their contexts.

By knowing where a device is located and what content is being consumed, marketers can provide messages and offers that are most likely to be of interest to the user. For example, if the user is in a restaurant district, surfing the Web for local eateries, and past behaviour indicates that this user enjoys a good pizza, a next-generation ad-hoc marketer might use this information to automatically send directions to a local pizzeria accompanied by a discount coupon.

## 10.10 A Decentralised Future and Boundless Opportunities

In the digital age, things change rapidly. New technologies enable better ways of communicating and computing, but at the same time can severely disrupt the old ways of doing things. The rise of ad-hoc and other alternative technologies will challenge existing business models, even as exciting new opportunities arise for new businesses and existing companies that are willing and able to adapt.

Large, highly successful enterprises have been built on the carrier model. When the Internet arrived, it became obvious that the greatest opportunities going forward lay in giving people more choice and control over the access they have and the content they share. New Internet companies sprang up almost overnight, and traditional carriers adapted by becoming Internet companies themselves.

Now, at the dawn of a new model for decentralised wireless broadband connectivity, ad-hoc technologies promise to place even more control into the hands of end users – moving the Internet model further toward the ideal of truly democratising information access. Ad-hoc wireless technology puts control into the hands of end users, who can use it to access all the information, services and communications they need, without relying on a particular carrier. It's a paradigm shift: instead of relying completely on a centralised provider, all your neighbours can now be a part of your communication system.

At the same time, ad-hoc technology enables more open business models that allow smaller companies and start-ups to get into the wireless broadband business. Peer-to-peer connectivity supports a larger and more integrated wireless ecosystem, with less need to rely on a centrally planned, controlled and maintained architecture. And for incumbent carriers who can adapt to take advantage of new opportunities, ad-hoc technology is the most flexible, cost-effective way to extend their existing, centralised architecture into new, decentralised wireless spaces.

Ad-hoc technology is a key enabler of the alternative wireless broadband networks that will extend the Internet to fulfil the promise of true anywhere/anytime/anything connectivity. Widespread adoption is inevitable, and is already beginning today – lowering costs, increasing functionality and opening up a world of future possibilities that we can only begin to imagine.

In this chapter, we have looked at a few of the opportunities nearest at hand. As further opportunities become apparent in the years to come, one thing is certain: ad-hoc technology will provide immense benefits to providers and users alike.

## Reference

[1] IDC, 'As mobile workforce grows, IT support could lag', *Computerworld*, 8 November 2005.

## Biographies

### Gary Grube, BSEE, MSEE, MBA

Gary is a Senior Motorola Fellow and directs all wireless research at Motorola Labs. Previously he held the Chief Technology Officer position at Motorola's Government and Enterprise Mobility Solutions business.

Gary joined Motorola in 1980. He has worked in the area of wireless systems development focusing on system architecture, intellectual property rights, and technology planning. Gary is a board member of the Motorola Science Advisory Board, and in 1994 he was named a Dan Noble Fellow, Motorola's highest award for technical achievement. He holds 107 issued US patents and has many more pending. A frequent public speaker, Gary has been called upon many times by the US Congress to testify as an expert in matters related to homeland security communications.

Gary serves as Vice Chairman of Safe America, a non-profit organisation focused on personal safety awareness and training. He co-chairs Operation Safe America which is connecting government and communities on homeland security awareness and preparation. In 2003, Gary was appointed by Mayor Richard M. Daley to serve on the Mayor's Council of Technology Advisors for the City of Chicago promoting high-tech around the Chicagoland area. He is also a member of the Executive Advisory Board of the International Engineering Consortium.

Gary earned a Bachelor of Science in Electrical Engineering (BSEE) degree at the University of Illinois, Champaign, a Masters of Science in Electrical Engineering (MSEE) degree from the Illinois Institute of Technology, Chicago, and he also holds a Masters in Business Administration (MBA) earned in the executive program at Northwestern University in Evanston Illinois.

Gary supports the technology side of Motorola's motor sports sponsorships where he oversees the application of current and emerging products at the track. His personal knowledge and interest in auto racing, combined with his experience in communications technology, helps Motorola leverage the challenging racing environment to test new technologies that can be deployed across other segments. He works directly with organisations like the General Motors Le Mans Corvette team and the Andretti–Green IRL Racing Team.

### Dr Hamid Ahmadi, B.S., M.S., PhD., FIEEE

Hamid has over 25 years of research, development and management experience in the areas of telecommunications, IP networking, wireless communications and network security. Named as Motorola's Chief Architect in January 2006, Hamid previously served as the Vice President of Technology and CTO in the Global Communications Sector at IBM Corporation. As lead technologist and CTO for the sector, he was responsible for formulating technical strategy for each of these industries by identifying new growth opportunities to enable customers to transform their IT and networking infrastructures using IBM's integrated solutions-based approach. Some key areas of focus included networking and services transformation for creating a unified architecture and service delivery platform for providing all services over Internet infrastructure, broadband wireless access (WiFi, WiMax and EVDO) technologies, digital content management and distribution, end-to-end-security architecture, IMS-based multimedia services solutions and a unified middleware platform for delivery and services integration.

Prior to rejoining IBM, Hamid was Vice President of the Global Network and IP Services Planning and Development organisations at AT&T Labs for five years. He is credited with having conceived the architecture and managed the development of AT&T Global IP network for all aspects of systems engineering, design, planning and services realisation supporting the needs of the network through service applications (layer one – layer 7) in support of security and managed services such as global VPN and MPLS offerings. His organisation was responsible for global development of AT&T's transport network, AT&T's IP backbone, global VPN, VoIP solutions and broadband access services. Hamid was also Vice President and Chief Technology Architect, Internet and Networking Systems.

Prior to joining AT&T, he was with IBM for 15 years, where he was the Director of Security and Networking Research at IBM T.J. Watson Research Center. His department there was responsible for research in the areas of communications systems, fast packet switching architecture, IP and data networks, mobile networking, cryptography, Internet security and security vulnerability analysis. He was responsible for research and development efforts that led to two IBM Wireless LAN products. He was also co-inventor of IBM Prizma Switch technology that has been deployed in several gigaswitch routers such as Celox Networks and Alcatel. In 1994, he was the recipient of IBM's Outstanding Innovation Award for Prizma Switch technology.

Hamid is an IEEE Fellow. He served on the Board of Directors of Packeteer®, Inc. (NASDAQ: PKTR), a leading provider of application performance infrastructure systems for four years. He also served on the advisory board of Airespace, a wireless switched-based company which was recently acquired by Cisco Networks. He is on the Industry Advisory Board of Electrical and Computer Engineering at the University of Maryland, College Park and UCSB. He was an Executive Board Member of the International Computer Science Institute at University of California, Berkeley. He served as founding editor-in-chief of the IEEE *Personal Communications* magazine, guest editor for the IEEE *JSAC* and was technical editor for the IEEE *Transactions on Communications*. He is also adjunct faculty at Polytechnic University of NY.

Hamid received his BS, MS and PhD degrees in electrical engineering from Columbia University, NY, in 1976, 1978 and 1983, respectively.

# 11

# Interference and Our Wireless Future

Dennis A. Roberson

## 11.1 Introduction

As with most things in life, the future of wireless communications is as heavily influenced by past actions as it is by ongoing innovations. In particular, the rich array of national and international laws and regulations which have been developed over the past century channel and in many cases limit the future directions for wireless communications. These regulations have clearly spurred the development, profitable deployment, and beneficial usage of radio technologies over this period. They have enabled us to ever more efficiently communicate, and have largely mitigated the huge interference issues that might otherwise have inhibited wireless exploitation.

At the same time, these historically helpful regulations now serve as significant encumbrances to the implementation of a range of exciting, emerging applications. These applications utilise ever more sophisticated advanced technologies that easily overcome most of the difficulties these regulations were meant to resolve. As we shall see, this has resulted in the unusual conundrum of regulated spectrum scarcity, where there is actually a great deal of spectral capacity available in time and space, but very little that is legally accessible. This circumstance is increasingly serving as a deterrent to the development and deployment of the new, highly desirable, and beneficial wireless applications that are being conceived in various research laboratories across the world. Later sections discuss this spectrum scarcity issue and the current regulatory efforts to resolve the issue. Presuming that this spectral scarcity issue can be resolved through the application of these approaches, the final section of the chapter describes some of the more interesting product and system developments and directions that should emerge over the next 25 years. The next section provides a short background on the critical regulatory developments over the past century.

*Wireless Communications: The Future*   William Webb
© 2007 John Wiley & Sons, Ltd

## 11.2 History

From the earliest days of commercial wireless telegraphy, interference mitigation and the often related challenge of frequency assignment to support solid wireless connection have been major issues. The ITU (then the International Telegraph Union and now the International Telecommunications Union), which had originally been formed to deal with wired telegraphy issues and standards, began its official regulatory efforts in 1903 with a meeting in Berlin attended by eight leading European nations and the USA [1]. This session took place only seven years after the invention of the wireless telegraph. The effort culminated in the first International Radiotelegraph Conference held in Berlin in 1906 and attended by 27 nations including representatives from 19 Central and Western European nations, four South American nations, Mexico, Japan, Persia (Iran), and the United States. This conference resulted in the adoption of the first widely agreed upon radio regulations which focused on ship-to-shore wireless telegraphy [2]. This conference was followed by a new 'International Radiotelegraphic Convention' which was signed in July of 1912 and established the requirement that all radio transmitters be licensed, thereby providing the mechanism for enforcing radio usage regulations.

In 1927, following the early introduction of broadcast radio by the Marconi Company in 1920, the International Radiotelegraph Conference met in Washington, DC, and for the first time allocated specific frequency bands for the various radio applications in use at that time. This included fixed, maritime and aeronautical mobile, broadcast, amateur and experimental bands. This was done to 'ensure greater efficiency of operation in view of the increase in the number of radiocommunications services and the technical peculiarities of each service' [3]. This did indeed enhance radio operation and facilitated a rapid rise in the deployment of radio applications which were rendered increasingly interference-immune based on the provision of areas of homogeneous spectral usage (bands with similar modulation schemes, power levels and the spatial allocation of one application per channel within a given band).

Also in 1927, in recognition of the broadening role radio was playing in society in the United States and the escalating challenges of interference between the services and the operators of similar services, the Federal Radio Commission was established as an independent agency with a direct reporting responsibility to the US Congress. This action transferred responsibility for radio frequency administration from the Bureau of Navigation within the Department of Commerce where it had resided since it was assigned the US enforcement role for the International Radio Convention of 1912. In 1934, the ill-funded and directed Federal Radio Commission was replaced by the properly chartered and reasonably well-funded Federal Communications Commission with the responsibility to foster the public good by the careful stewardship of their spectrum allocation role, wireless and ultimately all communications oversight. Of critical importance, this role encompassed both regulatory and enforcement responsibilities. Similar regulatory and enforcement organisations have been put in place in most countries and geographies in the world.

In 1947, the ITU and its frequency assignment process were strengthened by virtue of the ITU becoming part of the United Nations organization and through the formation of the International Frequency Registration Board (IFRB) within the ITU. This board has the increasingly complex task of guiding the use of the spectrum on a worldwide basis in an environment of national and local regulatory decision-making, and often competing and contradictory national and regional standardisation efforts and directions. This has created a maze of regulations and public and private administrative bodies that must be navigated to successfully deploy a global wireless application.

## 11.3 Spectrum Scarcity

In spite of the complexities and enormous challenges of spectrum allocation, regulation and enforcement at the national and international levels that have been briefly highlighted above, overall this process worked reasonably well from its formal adoption in 1927 until roughly the mid-1980s. By this time, unfortunately the easily accessible spectrum had been both allocated and the demand for spectrum to support new applications demonstrably exceeded the ability of regulators to re-allocate spectrum for these applications to use. The pace of development of new applications has continued to accelerate with the availability of first integrated circuit technology and specifically the microprocessor, and later digital signal processors and ever improving analogue radio components, and radio system knowledge.

The programmability of these devices has been particularly instrumental in providing platforms for the rapid development of more and more complex and capable access and modulation schemes. Similarly, the introduction and evolution of stack-based protocols has led to an ever improving means of controlling and more efficiently utilising the spectrum within each allocated application band. Together these advances have further accelerated the development of new and ever more exciting ways of utilising the spectrum for a broader and broader set of discrete applications and services. They have also dramatically increased the level of integration of various services in a single application band, and into discrete handheld devices. Beyond this, additional processors have been added to many devices and systems to enable even more exotic applications. See Figure 11.1 for a representation of the US frequency allocations courtesy of the US Department of Commerce, National Telecommunications and Information Administration, Office of Spectrum Management [4]. An up-to-date version of this spectrum chart is available online at www.ntia.doc.gov. It visibly illustrates the current allocated spectrum scarcity issue.

## 11.4 Regulatory Directions Toward Scarcity Amelioration

This spectrum conundrum has led to the establishment of three approaches to mitigate the spectrum scarcity issue. First, in the mid-1980s the ITU and the US FCC established the notion of 'unlicensed' bands to serve as the home for a heterogeneous set of applications in the broad category described as the Industrial, Scientific and Medical Band. Though initially established to take care of the specific needs of the areas included in its name, the band definition was expanded to provide the needed safety value for a wide variety of new networking applications including the extremely popular WiFi wireless local area network (WLAN) technology and BlueTooth, the most popular wireless personal area networking (WPAN) technology.

More generally, any application can use these bands without paying the often prohibitive auction cost for licensed spectral band usage. The only constraints are agreeing to abide by the specified power limit for the band, and being willing to share the band with whatever application may come along. This was a wonderful regulatory action unleashing the creative energies of numerous wireless application providers, and standards development consortiums. The downside in using these bands has been the corollary to the strength of the regulation,

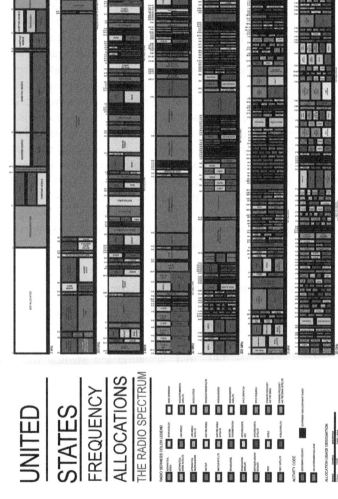

**Figure 11.1** US frequency allocation chart.

namely, since 'any application can use the band', an ever increasing number do use the band making the interference issue in the band ever more difficult. Beyond the wide variety of WiFi and BlueTooth deployments, this set of unlicensed spectral bands currently houses applications as diverse as microwave ovens, magnetic resonance medical imaging systems, street and overhead projector lamps, baby monitors, portable phones, proprietary 'last mile' technology, sensor networks (under the ZigBee Alliance) and industrial microwave processing equipment used in the production for such products as optical fibre cables, automobile glass, and no-wax floor tiles [5].

Based on the confluence of technologies in the existing unlicensed bands, we are now seeing the conditions emerge for the wireless 'perfect storm', or more colloquially, a 'quadruple whammy' created by the combination of (1) an increasing number of new application types, which are (2) selectively and in aggregate becoming more widely deployed, and which are (3) being more heavily utilized on an hours per day and days per week basis, and finally which are (4) being called upon to perform at ever higher data rates. Collectively this is creating a continuing radical increase in the amount of wireless information being transmitted, and the amount of spectral capacity being utilised. As a result of this increasing demand, additional unlicensed spectrum has been and is being carved out for wireless data networking applications. In the United States this 'new spectrum' falls under the Unlicensed National Information Infrastructure (U-NII) band structure. At this writing, both of the ISM and U-NII bands are rapidly filling up in the world's major urban financial and industrial centres.

Given the challenges with the unlicensed bands, regulators searched for other means of addressing spectrum scarcity. The second approach that has been deployed in the United States is the use of underlays in specific areas of the spectrum. In this case, the carrier must operate below the 'FCC Part 15' noise limit and must be implemented using a very broadband carrier (at least 500 MHz) to match the dictates of the regulations established in 2002 [6]. The implementations under this regulation are typically described under the generic title of ultra wideband based on the signal-carrying bandwidth used for this class of communications. The standardisation efforts in this space, under a group identified as IEEE 802.15.3a, has been one of the most negative aspects of the technology development effort in that two incompatible competing techniques were initially proposed with significant groups of companies quickly aligning with each. In early 2006, after five years of significant effort, both sides of the technology-based debate had become fully entrenched with neither garnering enough support to create a standard. This led to the unusual IEEE action of declaring that this standardisation effort was a failure and that the group's standardisation development authority was rescinded. Both UWB efforts are continuing to press forward towards the widespread commercialisation of this short-distance high-bandwidth communications technology, initially applying it to the needs in the consumer electronics application space. It is predicted that the marketplace will ultimately determine which of the two techniques will succeed.

The third approach to satisfying the need for additional spectrum would seem to offer the greatest potential, and at the same time has by far the greatest technical and regulatory challenges. This is the concept of spectrum overlays or the dynamic usage of previously allocated spectrum when and if the new, non-allocated user can prove that they will not disrupt the incumbent. The premise for this concept is the observation, described above, that while effectively all the spectrum in the frequency ranges of interest has been allocated, thereby firmly establishing spectral scarcity, most of the spectrum, in most of the places,

most of the time is completely unused! Figure 11.2 illustrates this point very well for two
of the USA's busiest cities – Chicago and New York.

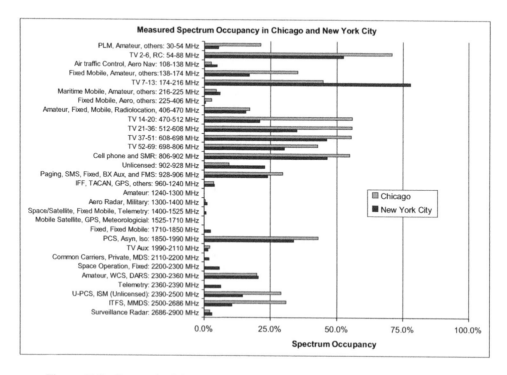

**Figure 11.2**   Bar graph of the spectrum occupancy in New York City and Chicago [6].

The net spectrum utilisation for the cities is 17.4% for Chicago and only 13.1% for New
York City. Obviously this suggests considerable opportunity for the deployment of overlay
solutions to satisfy the need for additional bandwidth for many years to come. The actual
means of implementing overlays is generally called 'cognitive radio' [7]. As this name would
suggest, the key capability of the radio is its ability to sense the spectral environment and
transmit only in areas of the spectrum that are not currently being used. In order to optimally
pursue this approach, the radio needs another crucially important technical characteristic,
namely the ability to transmit in a wide variety of spectral channels and ideally even a
variety of different spectral bands. For this reason, the technology is sometimes described
as 'frequency-agile radio', emphasising this aspect of the required capability.

Though the availability of spectrum is clear and the conceptual notion of utilising the time
and space dimensions more effectively to enhance the data carrying capacity of the spectrum
is well understood, the challenges of implementing this technology, such as overcoming the
hidden terminal problem (see Section 4.2.4), are extremely difficult. There are many other
issues to be dealt with, especially if the intention is to overlay an existing allocated area of
spectrum without disrupting the incumbent applications. Particularly vexing is the challenge of
quickly releasing a channel when the incumbent wants to use the spectrum. These difficulties are

also band-unique making the problem even more interesting if the intention is to be able to operate in a variety of bands. One set of spectral bands receiving a great deal of attention is the broadcast television bands. These bands consume a considerable amount of prime spectrum in the 54–806 MHz spectral range (note the large bands in the fourth and fifth rows of the spectrum chart depicted in Figure 11.1).

When viewed in a course-grained spectrum occupancy chart such as the one in Figure 11.3 from a recent spectrum occupancy study of Chicago [8], it appears that the television spectrum is completely utilised. In fact, because of the horizontal and vertical refresh aspects of historic scanning electron beam television tubes, a significant portion of the transmission time is not used to transmit data to the television and, at least hypothetically, could be used for other data transmission purposes. Because of the predictability of these retrace times, it is technically relatively trivial to use this spectrum for a secondary purpose. From a regulatory standpoint, however, this has been extremely challenging given the television station owners' strong desire to 'own' their allocated spectrum. This includes 'owning', controlling and profiting from any secondary commercial usage of the spectrum.

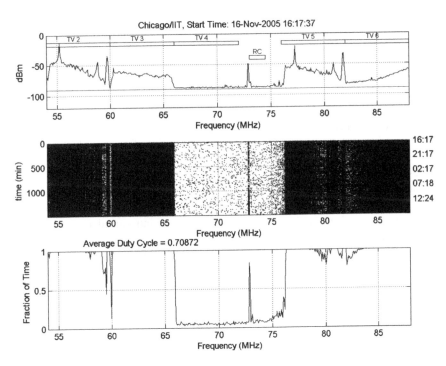

**Figure 11.3**   Spectrum occupancy for TV band, Chicago, 16 November 2005 [8].

The common denominator between all three of these spectrum scarcity amelioration approaches is the need to be able to handle interference in a more elegant manner than has historically been either required or possible. In particular, the whole premise of the historic spectral band allocation system has been to ensure that only homogeneous modulation schemes existed within bands, with guard bands in place to isolate each band from those on either side. In this environment, cross-channel and cross-band interference could be very well analysed and the implementation designed to minimise interference and therefore to maximise the application's robustness and/or throughput. Each of the three spectrum expansion techniques (unlicensed bands, underlays and overlays) creates an environment in which real or potential heterogeneous access and modulation schemes naturally exist in most bands and even in most channels. Specifically, in the unlicensed bands there are a virtually unlimited number of different modulation schemes that might be implemented. In the underlay environment there are far fewer possible modulation combinations to be dealt with in a specific band, but the number of bands that can be impacted by a specific ultra wideband signal is a totally new phenomenon to be dealt with. Finally, in the overlay case, the whole nature of the approach is to ensure that a new and likely quite different access and modulation scheme can operate without interfering or even having the incumbent know that the overlay signal exists.

## 11.5 Scarcity Amelioration Approaches

The technical means of dealing with these interference challenges logically sub-divide into receiver-based enhancements and transmitter-based improvements. While most of the historic focus has been on the management of the transmission side of the equation it has been increasingly recognised that by improving the noise immunity, selectivity and general sophistication of the receivers, the overall capacity and reliability of most wireless systems can be enormously improved. Beyond this, the new spectrum capacity enhancement techniques described above can be dramatically enhanced if the increase in interference implied by the deployment of these techniques has less impact on the receivers.

On the transmitter side, there are a wide variety of new interference reduction techniques that have been percolating in various university, government and industry research laboratories for the past decade or more. These include the continuing stream of ever better power management schemes. These increasingly dynamic techniques ensure that only the power needed for a quality connection is used to communicate to target receivers. These approaches have dramatically reduced the un-needed energy introduced as interference noise into the overall environment.

Beam-forming techniques focus the signal energy on the user station's unique location thereby reducing the interference related energy that would otherwise be sent to cover a broad quadrant of space. Similarly, multiple-input/multiple-output (MIMO) systems create spatial diversity enabling higher effective data rates in the same bandwidth, thereby improving the efficiency of the spectral use, the overall capacity of the system and the spatial–spectral capacity in general. Interestingly, this spatial diversity advantage is achieved to a considerable degree, even when the antennas are only separated by the width or the height of a cell phone. While we cannot create new spectrum, we clearly can dramatically enhance its utilisation.

## 11.6 Emerging Wireless Communications Devices and Systems

Frequent reference has been made to the large number of exciting, advanced applications that will require a vast increase in spectral capacity. It is now time to describe a few of these applications and more fundamentally the spectral usage directions that the applications represent. The first obvious trend is function integration at the device level. We all recognise that the mobile device 'formerly known as the cell phone' is now not just a high-tech phone with caller ID, missed call and placed call identification, single digit and voice dialling, and its own answering service. It is also a powerful, connected PDA (which itself is the integration of numerous historically separate functions – calendar, alarm clock, address/phone book, calculator, notepad etc.), an email point of presence, a text messaging system, a general-purpose Internet access point, a digital modem for your PC, a portable game machine, a reasonable resolution digital camera often with short duration video capabilities, an MP-3 music player, a pocket television, and soon a video player (the video iPod) etc. While this is both very exciting for some, and a little overwhelming for others, this source of greatly expanded cellular bit traffic is only the early harbinger of things to come.

Exciting developments are also occurring in the nomadic space with the extraordinary proliferation of WiFi (or IEEE 802.11). Beyond becoming ubiquitous on our university campuses, in coffee shops, hotels, transportation waiting areas, in our small-office/home-office (SOHO) environments, and increasingly within our homes, WiFi is now being deployed in public spaces in an ever increasing number of cities. At this writing this includes requests for information or bids or in some cases initial deployments in over 300 US cities including Boston, Chicago, Cleveland, Dayton, Las Vegas, Los Angeles, Milwaukee, New York, Philadelphia, Phoenix, Portland, Redmond, Sacramento, San Antonio, San Francisco and Tempe. The UK has a similar list of cities including Birmingham, Cardiff, Edinburgh, Glasgow, Leeds, Liverpool, London and Westminster with plans for many more. This trend continues around the globe with the following serving as examples of other well-known cities on the same path: Toronto and Waterloo (Canada), Jerusalem (Israel), Osaka (Japan), Auckland (New Zealand), and Taipei (Taiwan). Even the home of the first university in the western world, and of wireless technology (courtesy of Marconi), Bologna, Italy, is becoming an Internet-enabled city using WiFi coverage.

It doesn't end with wirelessly connecting cities either as I am personally engaged with a company that is providing wireless access to the occupants of boats moored in the slips at various marinas. Other companies provide wireless connections to camp grounds, and recreation vehicles parks. Still others are providing Internet access at truck stops and freeway oases. It seems that we will soon have the opportunity to obtain a broadband Internet connection via WiFi in almost any place we chose to gather. This will even include access on the airplanes, commuter trains and even buses that we travel on in addition to the current access available in the terminals where we wait for our public transportation.

Even as WiFi technology is becoming pervasive, a new technology called WiMax (IEEE 802.16), which promises to dramatically extend and selectively replace it, is emerging from the standards bodies and seeing early product deployment. Even before full WiMax compliant systems are widely available, WiMax-like wireless metropolitan area network (WMAN) broadband products and systems have been and are being widely deployed by wireless Internet service providers (WISPs) to provide homes, businesses, and private and public organisations with Internet access. These products and services will soon make the leap from fixed service offerings to full mobile services, providing broadband wireless Internet connections across the full range of Internet services. This includes significant provisions for supporting voice-over-Internet protocol (VoIP) services. This should ultimately create many of the same financial challenges for cell phone services that are currently being experienced by wireline-based phone companies.

As one would expect, all of this is radically increasing the availability and performance of the wireless Internet. Even as the volume of access points (APs) has increased, and the related connected users has swelled, the average broadband connection speed has also dramatically risen with the current IEEE 802.11n standard boosting the rated speed from 54 Mbits/s to a 270 Mbits/s data rate (actual data rate projections are from 100 to 150 Mbits/s). As a result, the total number of bits per second in this usage area has continued to swell. Application-wise, the usage of the wireless Internet is being driven not only by classical email, instant messaging, 'Googling', on-line shopping, and music and video downloads, but increasingly by mega-scale, multi-user action games with tens of thousands of players, and by creative multi-authoring activities like the online production of fanciful anime and manga (Japanese animated and paperback cartoons).

In addition, we see the rapid acceleration in the availability of BlueTooth devices ranging from cell phone headsets to keyboards, mice and printer connections, to wireless game controllers and connected personal multi-player game machines. As noted earlier, the emergence of UWB technology is enabling the wireless connection of immersive surround-sound entertainment systems which include large flat-screen display systems with associated high-performance speaker arrays supporting cable and broadcast HDTV, DVD players, MP-3 players, Internet-connected video game machines, AM/FM radio receivers etc. all connected wirelessly in your home. These integrated systems can be housed in a single room or distributed around the home, or commercial or institutional establishment.

Continuing the earlier application integration theme around 'the device formerly known as the cell phone', the various historically separate wireless connection schemes are also converging on these devices. Currently, cell phones include one or more of the following base cellular standards: GSM (usually with GPRS or EDGE for enhanced data handling), CDMA (with EV/DO and soon EV/DV), iDEN (WiDEN), CDMA/2000, or WCDMA (UMTS), with up to five or six frequency bands to provide connectivity. They also include GPS connectivity for location identification and navigation, and BlueTooth for headsets and other connections. Finally, they are now beginning to include a WiFi connection for in-building VoIP and other wireless Internet connections. This means that a single handheld device can have up to eight antennas (assuming one antenna per band), or even more if MIMO technology is used, and the ability to handle five or six unique modulation/multiplexing schemes. By historic radio standards, this level of complexity in a handheld device is nothing short of astonishing!

All of this is in our immediate future with widespread deployment for some of the more advanced features, capabilities and levels of integration extending over the next five years or so. Beyond this timeframe, more immersive technologies will begin to emerge. In particularly, the avatar-oriented multi-player gaming experience will both be dramatically enhanced and increasingly transferred to our normal day-by-day wireless communication experience. From a gaming standpoint this means a full three-dimensional capability and even the addition of various forms of haptic (touch) communications to dramatically add to the realism of the experience (think *The Matrix* without the uncomfortable and intrusive requirement to 'plug' your body into the system). This will lead to a 'just like being there' capability including full video and ultimately a realistic immersive virtual reality experience. Happily (and very unlike *The Matrix*), the virtual reality experience will be highly reflective of the true reality being experienced by the person with whom you are communicating. This will be extraordinarily handy for such activities as shared purchasing decisions with a spouse, a business partner, or a broker, or attendance at an important athletic, drama or music event for your child (or other close relative or friend) when you are unable to physically be in attendance. Given the global reach of the technology, this capability will even selectively enable shared travel experiences by families, or social and/or business organisations where only one of the member of the group is able to be involved in the physical journey.

As we move to 2025–2030, we will see the emergence of what today is very controversial technology enabling the ability to effectively communicate from 'mind to mind'. With a little practice, wireless technology (along with a considerable array of sophisticated sensor, processing and communicating hardware and software) will enable the capture of specific message-focused brain waves, transforming them into an easily modulated form and boosting the power to a transmittable level with a body-worn communication unit. This will enable desired information to be sent through a largely wireless system to the intended recipient. The system will enable high-fidelity communications in an efficient multimedia (pictures, videos, music, thoughts in general) fashion. The system will obviously need substantial robustness, very high bandwidth, and strong security features to ensure that only the intended recipient is provided with the transmitted information. This form of communication would clearly include provision for both business and personal exchanges, and over time it would be extended to serve multi-user communities, enabling meetings or other gatherings (lectures, workshops, concerts, advanced games etc.) to occur at the thought level.

Beyond 'human to human' exchanges, this advanced wireless system capability will also provide an individual with the ability to communicate with the various electromechanical 'things' we require or desire to support us in our daily affairs. Specifically this would include personal transportation systems, home and office heating/cooling systems, appliances, consumer electronic devices and systems etc. and the new devices that only exist to any significant degree in the world of science fiction. In particular, it is anticipated that the long awaited arrival of the first crude humanoid 'servant' robots would occur in roughly this timeframe. Wireless communications will play an indispensable role in controlling, directing and otherwise interacting with these helpful 'automaton friends'.

These musings on our wireless future have likely been extended to the limits of utility for this book. It is important, however, to return to the fundamental thesis of this chapter, namely, that managing the spectral utilisation to dramatically enhance the communications/data carrying capacity and robustness will likely be the most critical element needed to enable this rich and exciting future. If we are to achieve this vision (or one anything like it) we will need to rapidly and radically improve our understanding of the static and dynamic characteristics of the interference environment. Using this knowledge, we must proceed with the continuing provision of far-sighted spectrum policy enhancements to support the development and deployment of the kinds of exciting and beneficial products and systems highlighted in this section.

## References

[1] Captain L S Howeth, USN (Retired), *History of Communications: Electronics in the United States Navy*, pp 547–548, 1963.
[2] *International Wireless Telegraph Convention (Convention Radiotélégraphique Internationale)*, issued by the Government Printing Office in Washington, DC, for the Navy Department's Bureau of Equipment [N5.2:W74/6], 1907.
[3] www.itu.int/aboutitu/overview/history.html
[4] US Department of Commerce – National Telecommunications and Infrastructure Administration, Office of Spectrum Management, *United States Frequency Allocations – The Radio Spectrum*, October 2003.
[5] Bennett Z Kobb, *Wireless SPECTRUM Finder, Telecommunications, Government, and Scientific Radio Frequency Allocations in the US, 30MHz to 300GHz*, McGraw-Hill Telecom, 2001.

[6]  FCC, First Report and Order 02–48. February 2002.

[7]  Joseph Mitolla III, *Cognitive Radio: An Integrated Agent Architecture for Software Defined Radio*, Royal Institute of Technology (KTH), Stockholm, 8 May 2000 (ISSN 1403–5286).

[8]  Mark A McHenry, Dan McCloskey, Dennis A Roberson and John T MacDonald, *Spectrum Occupancy Measurements – Chicago, Illinois – November 16–18, 2005*, 20 December 2005.

## Biography

### *Dennis A. Roberson*

Dennis Roberson is Vice Provost and Executive Director of the Institute of Business and Interprofessional Studies, provides oversight for the Jules F. Knapp Entrepreneurship Center, is the Principal Investigator for a US National Science Foundation funded Wireless Research Laboratory, and serves as a Research Professor of Computer Science, all at the Illinois Institute of Technology. In these capacities, he is responsible for IIT's undergraduate business and related co-curricular programs, and its concentration on entrepreneurship, leadership and technology-based projects. He is also responsible for working across the university to assist in technology transfer, and the development of new research centres and technology-based business ventures.

Prior to IIT, he was Executive Vice President and Chief Technology Officer of Motorola. Professor Roberson has an extensive corporate career including major business and technology responsibilities at IBM, DEC (now part of HP), AT&T and NCR. He has been involved in leadership roles in a wide variety of technology, cultural, business and youth organisations and serves on several public and private boards. He is a frequent speaker at technology-related meetings and conferences around the globe.

Professor Roberson has undergraduate degrees in Electrical Engineering and in Physics from Washington State University and an MSEE from Stanford.

# 12

# Three Ages of Future Wireless Communications

Simon Saunders

## 12.1 Introduction

I was initially delighted to be invited to contribute to this book amidst such distinguished company. I then sat down to write my contribution, and briefly wondered whether the editor had perhaps first approached a long succession of potential authors, each of whom, in turn, had wisely declined, while the editor mined his list of contacts ever deeper before reaching as far down the list as me. For writing about the future is truly a poisoned chalice. The writing is preserved in the annals of distinguished libraries and institutions, for all to see and smile at in years to come. Many are the articles debunking the budding futurologist, whose predictions of jet packs and teleportation in the imminent future fail to appear. Here we are, well into the first decade of the millennium, and the shiny suits and intergalactic warp drives beloved of twentieth-century futurologists have singularly failed.

I was briefly heartened by the fact that most books about the future fail to be read long after their publication, when I remembered reading the editor's article analysing the predictions from the first edition of this book. So there is no escape and no chance to appeal to a future collective amnesia. Down to it then.

In order to provide some organisation to this chapter, I have first divided the next 20 years into three fairly arbitrary periods of five years, five years and ten years. For reasons which will become clear, I have given these names as follows:

- 2007–2011: The age of wireless proliferation
- 2012–2016: The age of wireless similarity
- 2017–2026: The age of wireless mundanity.

*Wireless Communications: The Future*   William Webb
© 2007 John Wiley & Sons, Ltd

The third age spans the longest period simply due to the cloudiness of my crystal ball and a desire to stand at least some chance of saying 'the dates were only indicative' when readers challenge me to explain why their shares in wireless warp drives have not realised their expected value.

In each age I have divided my thoughts regarding the future of wireless into the following general areas:

- *Services and applications*: what people are doing with wireless.
- *Devices*: what people are using to access their wireless services – the user's means of access to the wireless world.
- *Infrastructure*: the changing equipment and system topologies for providing users with wireless access and interconnection to other systems and data sources.
- *Air interfaces*: the 'over the air' languages and protocols via which devices and networks intercommunicate.
- *Spectrum*: the electromagnetic medium which permits wireless to work in the first place.

The first four of these follow a fairly conventional communications hierarchy. The last deserves a special mention, however. Spectrum, in the context of wireless communications, acts as the 'layer zero' – the very stuff of which wireless consists. Without it, wireless would be meaningless. A full understanding of the likely trends and outcomes for wireless requires knowledge of the detailed nature of spectrum – how it interacts and how legislation and regulation will increasingly permit its exploitation to an extent which may currently appear absurd. The fundamental characteristics of spectrum are worthy of mention here.

### Spectrum is finite

For practical communications purposes, frequencies from a few kilohertz to perhaps one hundred gigahertz represent the entire 'universe' and all wireless applications inhabit this space. In practice, the prime territory – the 'sweet spot' – is between a few hundred megahertz and a few gigahertz, where a reasonable balance between antenna size, range, bandwidth and system cost exists. Although the upper useful limit may extend somewhat due to reducing technology costs, this range has to provide the basis for almost all current and future applications of wireless. Although this book addresses primarily the communications applications, there are many other industrial, scientific and medical applications that also occupy the same space. I will make clear in this chapter that I do not believe this creates an ultimate limit on the potential for wireless, but it does represent a challenge of a fundamentally different kind to a wired environment, where adding more bandwidth is simply a case of adding more wires. Ingenious ways of reducing the bandwidth occupied by signals supporting given applications have been and are prime drivers of technological innovation in the past and near future of wireless. For the longer term, I will argue to the contrary: it is ingenious ways of filling *more* spectrum that will be key to producing an unconstrained wireless future.

### Spectrum is shared

All users share the radio spectrum, so interference is always a possibility. This is even the case when devices are intended to occupy and respond to different bands, because there are numerous interference mechanisms that can cause this, including spurious emissions and responses in transmitters and receivers, adjacent channel (and beyond) interference and nonlinear effects such as intermodulation. These mechanisms all become most significant when communication devices are numerous, when they are operated close together and when they are designed to have a low cost, all of which are growing characteristics of tomorrow's wireless systems. On the other hand, these mechanisms are also at their most damaging when systems are high-powered and use large, high-gain antennas, trends away from which most wireless is moving. This gives scope for optimism that interference chaos can be avoided in the future. Nevertheless, appropriate management techniques for multiple systems in shared spaces is a critical success factor for realising reliable wireless systems, and I will argue that this process will increasingly become an embedded and automatic function.

### Spectrum is complex

Spectrum complexity manifests itself in many ways. Regulations and legislation currently allocate the spectrum into bands, with myriad channel assignments and associated technical conditions and restrictions on use. The electromagnetic characteristics of radio waves are difficult to understand, and radio-wave propagation laws in particular lead to many uncertainties as to the range of coverage and interference produced by individual systems and as to the performance that will be experienced. All of these effects vary with time and location. As with the need for automated management processes, this also means that systems must become increasingly self-aware in order to realise their potential. Users and even the professionals who deploy wireless networks need to be supported by automated technology that allows them to work without understanding – or often even being aware of – the arcane details of wave propagation. Indeed, such smart devices will actually welcome complex radio environments since these produce opportunities for spatial reuse and consequent capacity and bit rate increases. Such automation also provides mitigation for a near-future growth limitation: there simply are not enough RF engineers to scale today's radio planning processes to support future needs.

### A word on tenses

In a book on future matters, it is inevitable that the future tense will gain more than its normal share of page real-estate. To avoid this becoming tiresome, I have avoided it almost entirely and have written about each age as though it were being experienced by a technically savvy user at the end of the age, rather as a science-fiction writer would. As well as breaking up the monotony of 'will' and 'may be', I hope this also helps the reader to imagine how it might feel to experience these wireless futures.

The three wireless ages are now described in turn. The final section of the chapter provides a summary of the main predictions.


## 12.2 The Age of Wireless Proliferation: 2007 to 2011

### 12.2.1 Introduction

It is now late 2011. The last five years of wireless can best be characterised as simply being about more of everything – more operators, more frequency bands, more networks, more air interface technologies, more devices and especially greater volumes of data being carried wirelessly. This proliferation has led to considerable challenges and rapid innovations, but not at the expense of any slowdown in the growth of wireless.

## 12.2.2 Services and Applications

After many years of searching for the 'killer applications' that would drive data use, the marketing departments of major operators for the most part gave up trying to second-guess user behaviour and allowed users and start-up enterprises to build services and applications which pleased themselves, rather than the network operators. As a result, data volumes and needs exploded, so data became a genuinely essential component of operator revenues and services in a way that had never previously been the case, voice having previously been of prime importance. Increasingly these services are created using the same technologies as in the fixed Internet, where web services are creating a 'MobileWeb2.0'. Such approaches are well suited to wireless connectivity, as they are optimised for connectivity of variable bandwidth and provide hosted storage of application data.

Voice revenues per user have been driven downwards via competition, but voice services, often via packages of large volumes included at fixed price, act as a key driver for customer retention. Overall revenues per user have increased due to the take-up of data services.

The first indication of serious data proliferation was the very widespread use of 'push' email services. These services were increasingly built-in functions of devices and email servers, increasing the availability well beyond the limited community which had experienced these services in the years prior to 2007. Deploying such services (e.g. the popular 'Blackberry' devices) had previously required a policy decision by corporate IT managers and a substantial subscription-based investment, while the new generation of devices made it a simple feature to be enabled. As well as reaching a much wider business community, hosted email services were offered widely by both wireless operators and by Internet service providers, expanding the use by consumers.

Once push email was in the hands of young consumers, a viral effect occurred, in which mobile email started to be used in much the way in which text messages previously were, with users encouraged to send more messages in the expectation that they would reach their destinations almost immediately. While MMS had provided the promise of sending pictures and video clips, the complexity of configuration and the need to find users with compatible phones meant that it never really took off. Instead, the ease of attaching pictures and clips to emails which could be read from both mobile and fixed devices led to a very rapid growth in data volumes. Today, SMS as a medium for mobile text has passed its peak in the most advanced markets and mobile email is increasingly the main choice.

In parallel with this, and growing almost as rapidly, were mobile presence services, which extended the popular instant messaging services such as MSN Messenger, Yahoo Messenger and the like across mobile devices. Knowing who in your contacts group was online made it all the more compelling to send emails, instant messages and pictures, leading to an enormously rapid increase in the associated use. Billing models were problematic, with users initially put off from enabling presence services as they were concerned about large data bills. As a result, many networks chose to stop billing for the traffic involved in updating the lists of online contacts, leading to a rapid increase in usage when users were alerted continuously as to available users and used that to decide which users they could interact with immediately. This also drove intensive use of wirelessly enabled social networking sites, which became one of the prime mobile destination sites. Such sites allowed users to opt-in to groups of users in the locality or with similar interests, greatly expanding the size and scope of their personal networks.

In the enterprise environment, it has taken longer than expected for wireless PBX services to be widely adopted. The initial promise was that mobile bills could be substantially reduced by using dual mode phones that would use the corporate wireless LAN network whenever in range via multi-mode phones. Several factors slowed this down: lots of W-LANs were simply not up to the job, leading to dreadful coverage, availability and call quality and the phones were both expensive and had poor battery life.

Cellular network operators also took away many of the potential advantages of wireless PBX services. Once such systems became widely available they responded by providing 'on net' calls amongst users in one corporate account via unmetered tariffs, and by including 'friends & family' style deals, where the most commonly called phones for each individual corporate user were billed at a rate comparable with fixed lines anyway. These issues are only recently being resolved, with patchy but significant deployment of voice over Wi-Fi phones in enterprise environments and communication suites produced by major software vendors rather than by traditional vendors of voice infrastructure. Typically these systems are provided as part of a deal with an operator to ensure proper seamless mobility between networks. The main drivers for this are improvement of indoor coverage and provision of increased data rates and capacity, rather than call cost reduction.

A major problem is occurring with the mixing of personal and corporate wireless data applications. Personal subscriptions to music and other entertainment services produce data volumes that can far exceed those required for corporate use, yet it is difficult for companies to separate these uses without preventing valid business applications. This is analogous to the issues experienced previously with private fixed Internet usage, but more severe as the financial implications are much larger.

Several niche operators have created services to provide fixed wireless broadband connectivity to businesses in competition to established fixed-line carriers. These are mostly focused on small-to-medium enterprises and operate in city centres and in specific business parks, often offered in collaboration with the landlord of the shared business space. There are also several operators providing mobile data over Wi-Fi and WiMax outside of the main cellular operators, but take-up for these networks is limited due to device restrictions and the limited coverage. Such networks have served, however, to modify the ways in which users are charged for services, with charges for stand-alone access (e.g. Wi-Fi hotspot access to the Internet) becoming rare. Instead, users are either charged for packages of services they can access, or services are cross-subsidised via advertising, with access being free to the user, depending on the network provider.

The domestic environment has seen significant changes. Service providers now provide or recommend a specific range of supported routers, which provide a mix of services including wireless voice, video on demand, television and Internet connectivity. These are complemented by a range of add-on services, such as home security alarm monitoring, including CCTV and IT security. Such routers give service providers a way of reducing churn and providing services that increase margins well beyond those available from the highly competitive market for undifferentiated Internet access. The wireless technology in use varies, including Wi-Fi, 2G and 3G operation within the router.

Such services also create a linkage between fixed and mobile operators, since the choice of service provider at home limits the operators that users can access in the wide area outside of the home. As a result, several strategic alliances have been created between the major fixed and mobile operators, with the possibility of closer ties in the future. One driver for this was the increasing leadership shown by mobile operators in the consumer market, while fixed operators have increasingly dominated in the business environment.

These domestic trends are starting to drive changes in the business market, just as domestic operation of Wi-Fi access points drove the adoption of enterprise Wi-Fi. Wide-scale domestic adoption also drives lower costs, improving the business case for enterprise adoption, although typically there is significant challenge in doing so. Domestic-quality access points are not usually appropriate for enterprise use, particularly in terms of security and management processes.

There is strong take-up of mobile TV services by consumers. This has driven considerable extra revenue and value for mobile operators, since users pay a monthly subscription that provides the operators with long-term predictability and a typical increase of ARPU for participating users of 10–15% depending on the service package. Other bandwidth-hungry mobile services such as mobile multiplayer gaming are also showing significant increases. Mobile video telephony is available on several networks, but has failed to be taken up in any volume.

In developing nations or those with large areas of sparse population density, 3G technology is being used to provide an initial level of both voice and broadband connectivity, often supplemented with 3G/Wi-Fi routers to extend connectivity to many devices.

## 12.2.3 Devices

The growth of large numbers of data applications has enormously increased the range of form factors for mobile devices. PDAs that do not provide several forms of wireless connectivity are redundant, as users are not prepared to tolerate a two-box solution. A PDA with proper integrated wireless connectivity produces a multiplicative, rather than additive set of synergies between the respective devices. However, this has meant that the range of options of technologies available within devices has become even more confusing. The only certain way of ensuring that your new device works with the range of services you are interested in is to buy it as part of a service package from your preferred operator.

Smartphones are probably the most common basic device, with a form factor similar to conventional phones and with small integrated keyboards. They have significant solid-state capacity (typically around 10 GB of flash memory) and include music and video players. Increasingly there is a strong market for mid-size devices. Some of these are consumer-type devices for personal communications and entertainment, which evolved upwards from portable gaming consoles with embedded Wi-Fi and cellular technologies. Others evolved downwards from laptops, catering for the majority of business users who simply need access to office applications, web browsing and calendar functionality rather than high-performance computing. In many cases these are also being used to wirelessly access web-enabled applications such as online collaboration sites, location services and online versions of office-type applications. Such devices perform all of these functions, while having hugely extended battery life and excellent portability and connectivity compared to full-sized laptops.

These devices, and others that evolved from portable MP3 and video players, share something in common: large amounts of local storage. Some achieve this via miniaturised hard disk drives, but storage of over 200 gigabytes is now available in solid-state form. Users of these devices synchronise all of their music, plus stored TV programmes and other video content. New music and video and incremental updates to subscribed programmes take place wirelessly. These services started as free podcasts, but are increasingly the main mode of receiving personalised broadcasting content.

Outside of the major industrialised nations, there has been considerable growth in the use of ultra-low-cost phones, helping to increase mobile penetration in developing parts of the world, often to places where the penetration of fixed phones has never exceeded a small percentage of the population.

## 12.2.4 Infrastructure

In order to service the increased data demands, the 'depth' of 3G coverage is being extended. The number of 3G macro-cells has grown steadily to around twice the number of cells deployed in 2007. City centres have seen deployment of street furniture-mounted distributed antennas systems to increase outdoor coverage, with some extension into high-traffic corporate and public buildings. In-building systems in large-scale buildings are increasingly being deployed using 'distributed base stations' or digital distributed antenna systems that distribute coverage to multiple radio heads digitally via gigabit Ethernet LANs. Conventional distributed antenna systems that distribute signals in analogue form to multiple antennas are still very much in evidence in mid- and large-scale buildings. Increasingly these offer a very wide bandwidth in order to provide for the wide range of frequency bands and technologies. Pricing for digital systems has, however, reduced significantly due to the continued growth of high-speed computation power and increased manufacturing volumes.

An alternative approach to indoor coverage is via small base stations. Mirroring previous trends in W-LAN, this started in the domestic environment and has grown towards large enterprise deployments. Domestic 'femto' cells were launched by some networks in late 2007, allowing users to purchase a base station with a package of home services, thus relieving capacity from the macro-cells while getting users to finance them. These have now reached the point that several Internet service providers have made deals with the cellular operators to provide these as standard embedded functions in home routers. Enterprise variants of these products are now being deployed for mid-range environments, where networks of these base stations are connected across a LAN and managed by the operator.

Many limited deployments of so-called 'wireless cities' were created, consisting mostly of Wi-Fi access points deployed on street furniture (lamp-posts, CCTV poles etc.). These provided an initial boost to some Wi-Fi service providers, but several of the bigger fixed-line operators, who were primarily interested in displacing voice revenues, found it more economical and reliable to provide such services over cellular technologies once they obtained licences to do so. The mesh technologies used to interconnect them are not suitable for high rate services, and the deployments have concentrated on city centres only. As a result, these deployments have mostly been used to support services for local authorities and for opportunistic access by Wi-Fi users, but have not created an alternative to wide area mobile networks.

## 12.2.5 Air Interfaces

There have been many incremental changes to existing air interfaces. 3G networks have been upgraded to include high-speed data extensions (HSDPA and HSUPA) as standard, as these were found to be an essential to realise a data experience acceptable to users used to wired broadband.

As a result, WiMax is not seeing as rapid a take-up as expected, although there are some new entrant operators who have built out WiMax from scratch. These operators are reportedly highly frustrated by the ongoing delays in mobile WiMax, including interworking, mobility and handset availability. Nevertheless, plenty of mobile computing devices do include WiMax capabilities and this provides an alternative path to Wi-Fi for some users. WiMax has found its major success in fixed access and backhaul applications.

Ultra wideband technology briefly attempted to become a LAN technology in competition with Wi-Fi, but quickly became consigned to a personal area (PAN) technology, just as BlueTooth had done in the early 2000s. This by no means indicates a lack of success; indeed in its guise of Wireless USB it is quickly becoming ubiquitous for new computers and peripherals. This was mainly due to the enormous installed base of Wi-Fi devices, but also because extended variants of Wi-Fi (IEEE802.11n and beyond) exceeded the range and data rate available to levels at or beyond those available from UWB over medium distances (tens of metres).

Fast Wi-Fi systems are being used extensively for domestic applications such as home video redistribution, which enables, for example, multiple TV sets to receive digital TV even when the coverage directly from digital TV transmitters is insufficient. TV sets commonly now have integrated Wi-Fi capabilities, allowing them to be used for both conventional TV applications and as computing monitors.

Complex domestic products have become easier to use as a result of wireless technology. Systems are no longer wired up. They are placed in proximity and connect interpretatively over UWB and Wi-Fi. Range and room boundaries define the natural interoperation partition. Multi-space appliances rely on LAN technologies and service advanced discovery mechanisms and appear as virtual controls on relevant appliances.

For mobile TV applications, DMB and MediaFlo are reducing in popularity as DVB-H advances, but networks for the latter are still not widely deployed and used, partly because of spectrum availability in different countries and partly because operators are reluctant to divide their new source of enhanced ARPU with providers of DVB-H networks. Devices are also still expensive, so 3G is more commonly used, producing a major driver for 3G network upgrade, particularly for indoor use. After an initial period of being offered using standard 3G, these are now being deployed using 3G-broadcast and multicast services via MBMS (Multimedia Broadcast and Multimedia Services) to minimise the network load. Since the capacity available from such services is still somewhat limited, however, most content is downloaded from a rotating 'carousel' of content, catering for a wide range of listening and viewing tastes and stored locally. Users then select from a set of standard and personalised 'channels' which are synthesised on local devices according to their preferences, with only the most time-sensitive content actually received in real time.

## 12.2.6 Spectrum

There has been progressive opening up and liberalisation of several key bands. This has caused a wide range of newer operators to appear providing mobile services, often those which were formerly restricted to fixed line services. Also several ISPs and consumer brands (including a major supermarket chain) have become real operators in themselves. Instead of incurring the large capital cost of macro networks, however, they have created an alliance of roaming partners, and have built small and medium area networks at key locations for their customers (including at all branches of the supermarket chain) and permit users to access services at all locations within the alliance. As a result there has been considerable consolidation of the mobile cellular market, with 90% of wide area mobile services now being provided by the 'big three' international operator groups.

Spectrum liberalisation has also enabled existing operators to utilise 3G technologies in bands formerly reserved for 2G, giving them greater flexibility in delivering high-rate services economically into rural areas, to deliver broadcast services without excessive reduction of capacity for point-to-point applications and to relieve congestion in traffic hotspots.

Considerable spectrum congestion has been encountered in the popular 2.4 GHz band. Many enterprises operating in multi-tenanted office buildings attempted to operate voice over Wi-Fi systems and found the interference caused unacceptable disruption to operation. This was not apparent when they had previously used their systems for data applications only, for even high-rate applications benefit from statistical multiplexing gains when sharing a channel and users are usually unable to distinguish interference effects from those blameable on some distant application server or on the Internet generally. Although the 5 GHz band is increasingly used, with far less such problems, for enterprise applications, most consumer portable devices and many legacy devices do not operate on this band. This has been a major contributory factor to the relatively slow adoption of Wi-Fi for wireless PBX services.

In 2010, the sunspot cycle peaked at an unusually high monthly average sunspot number of 180, with several solar storms well beyond this number [1]. The associated high levels of solar radiation reaching the Earth caused widespread disruption to major networks over many frequencies right across the Earth. The consequences for the global economy in the short term were very significant and there is a United Nations project under way to establish a global strategy to avoid a recurrence of such events, either naturally or as a result of some act of terrorism.

## 12.3 The Age of Wireless Similarity: 2012 to 2016

### 12.3.1 Introduction

It is now late 2016. The key changes over the last five years have mainly been about the coming together of the many variants of wireless systems into an increasingly coherent global whole. Different systems provide intense and often seamless interworking. As a result the divisions between operators of fixed and mobile technology and between the associated infrastructures have eroded significantly. The complex mix of technologies has become less relevant as we now have devices and networks that span multiple technologies and which make the transitions between these technologies mostly invisible to the user. Even for the operator, whose infrastructure is increasingly 'all-IP', the management and provisioning of services hides them from the detail of specific wireless technology choices to a large degree.

### 12.3.2 Services and Applications

In 2015, data volumes associated with wireless machine-to-machine applications exceeded those between people. This was not widely anticipated, since personal data volumes were increasing very rapidly, but finally the EC and US laws relating to fair trade required that every item sold had an RFID tag embedded and the immense number of low rate transactions in tracking all new items sold created the necessary momentum. Much of this traffic was over local networks, particularly in shops and warehouses, but wide-area mobile networks were increasingly used to ensure that such services were available almost everywhere.

Following the switchover of TV broadcasting from analogue to digital in many parts of the world, mobile TV has become an essential and standard part of consumer behaviour. Based on past choices and expressed preferences, individualised channels are created and consumed, with the opportunity to recommend individual programmes enjoyed to other users with similar profiles and to create mixes and blends of content individually personalised, all from a mobile device. In many cases these services were initially offered by mobile operators as an independent service, but are now tending to be part of an add-on package from TV service providers where users can pay extra to see all, or a selection of, the channels that they already subscribe to at home. They also have options to extend their service to include higher definition or 3D viewing options.

Wireless systems for control of lighting and heating systems in major industrial buildings are becoming common, although by no means mass market yet. These are driven particularly by high energy prices and punitive taxation for energy wastage. Domestic systems are also being offered and installed, although competing standards are limiting growth. This is fuelled by the use of wireless to avoid complexity in configuration.

In the domestic environment, wireless also now enables a wide range of entertainment and information access services. For example, favoured photographs follow the user, devices are powered up according to user behaviour, information displayed depends on the location of the device, and content can be spread across multiple devices.

Presence services are now complemented by location services. These were very slow to start but location is now a standard enabler for many mobile applications.

## 12.3.3 Devices

The rise of near-field technologies has permitted users to make use of input devices wherever they find them. Security is overcome using biometric scanning. A user need simply reach for the nearest keyboard and select a screen, and is able to input and output data without a problem.

Devices increasingly include several antennas for MIMO operation in a sufficiently compact form to be unobtrusive and inexpensive.

Domestic routers are no longer separate from home PCs: instead they contain all the computing power and storage needed for the whole household and run domestic variants of standard operating systems from the major OS vendors. These are accessed via 'thin client' technology, extending desktops and displays over many devices throughout the home according to the user's needs.

## 12.3.4 Infrastructure

There has been a significant evolution in the structure of wireless operators and their relationship to the operation of their infrastructure.

Operators rely fundamentally on three major elements:

- user accounts and the underlying credit relationships and knowledge of past usage and expressed preferences

- a network, consisting of both core and radio elements, over which to transport traffic to cell sites and to end users
- licences to operate, covering both the provision of services and particularly the spectrum used.

In the past, operators worked on a model that required all three elements in almost every case. Today, however, operators increasingly recognise that the only essential element is their relationship with their customers. Their knowledge of and response to their users' requirements is fundamental to their value. The network represents a major capital asset, with a return on investment over a long term. This asset requires substantial ongoing expenditure on maintenance and upgrades, which often cannot be adjusted sufficiently rapidly to match to changing services and customer requirements. Many operators therefore forego ownership of a network in favour of a wholesale service provided by an aggregator, providing the operator with increased certainty of expenditure and improved return on capital.

Although a licence to provide services is required by the operator, licences for spectrum tend not to provide significant differentiation between operators, and can sometimes act as a restriction on flexibility in adopting new technologies or devices to provide services. Also, complying with licence conditions requires detailed knowledge of the underlying network, so often operators are providing services using spectrum licences owned by the wholesale network aggregator. Many exceptions exist where operators still insist strategically on controlling all three elements, but there is a clear trend towards separation of these. This also enables more operators to enter the market, continuing the trend that started with the growth of mobile virtual network operators (MVNOs). It has also led to the growth of specialist spectrum brokers who have only spectrum assets, but no network or customers.

Following the sunspot calamity of 2010, a concerted international project led to the launch of three separate new satellite networks, two based on low Earth-orbiting satellites and the other using a complex network of highly elliptical orbits. These provide resilient global broadband wireless access as a fallback, to ensure continuity of the global economy in extreme circumstances. They are in operation continuously, however, so that there is now nowhere in the world outdoors where relatively high-rate reliable mobile network access cannot be obtained from at least one of these networks. This has provided infill to areas of the world where terrestrial networks are not economical, helping improve access to technology for less developed locations.

Backhaul from base stations has become highly congested in many key areas and acts as the ultimate limitation on capacity growth in many cases. This occurs because the backhaul links points of aggregation which set overall service quality for populations of users rather than individuals. Wireless backhaul is increasingly common over a variety of frequency bands, often using equipment and frequencies intended usually for user links that are occasionally under-utilised to ease congestion peaks.

## 12.3.5 Air Interfaces

It is increasingly apparent – and more importantly, well accepted – that air interfaces share the same basic trade-offs between range, coverage, capacity and performance, although the specific parameters vary. In general it is difficult to find a clear dividing line now between the multi-carrier CDMA systems which evolved from 3G cellular and OFDM systems, including Wi-Fi, WiMax and UWB. These are all part of one interlocking set of standards following some intense standardisation activity – including some acrimonious debates regarding cross-licensing of relevant patents and other intellectual property rights. The flexibility of today's equipment and the desire to avoid such delays has also forced a change to a more IT-like predominance of de-facto standards driven by commercial rather than regulatory forces.

Instead of the creation of more basic air interfaces, there has been increasing standardisation of 'meta' air interfaces that provide a basis for devices of varying degrees of sophistication to share spectrum. The most intelligent devices shift frequency, time, code and spatial channel according to their needs for bit rate, resilience and latency. Even in basic devices, the same technologies can now be accessed over a wider range of frequencies. 3G, for example, is now available in around twenty different bands worldwide. Older devices may not have this degree of flexibility, but networks increasingly take this into account when allocating radio resources to ensure that such devices do not 'hog' radio resources. These 'meta' air interfaces represent the evolution of the concepts of cognitive radio and related technologies which were first discussed in the early part of the millennium.

Exceptions to this unification include low power consumption and high energy efficiency per-bit technologies, evolved from Zigbee and other technologies, which are intended for shorter range low-rate applications such as for domestic and industrial automation and building control. This arises since their needs differ greatly from those of the high-rate-capable air interfaces for mobile devices.

Standardisation work continues to ensure that such mechanisms enable all the necessary flexibility, without prejudging the technical mechanisms used to achieve them.

## 12.3.6 Spectrum

Spectrum availability has become a real problem. In certain key bands – including those in the bands predominantly used for mobile wide area networks – spectrum trading relaxation led to the creation of several organisations who were acting solely as spectrum brokers, with no infrastructure or end users as assets. These organisations recognised the associated opportunities for large returns on capital and an increasingly dynamic and liquid market was the result.

In early 2016, however, two young spectrum traders in the UK found a way to hoard large numbers of spectral slots without anyone noticing and then proceeded to act as virtual monopoly suppliers for several key bands in particular high-traffic areas. In principle there was nothing illegal about this, but it led to the prices of spectrum exceeding the value of the services carried over that spectrum. Several major corporations were prevented from delivering their goods and services economically as they had no fallback arrangements in place. Their share prices fell rapidly, followed several weeks later by a major collapse in the UK stock market as analysts revealed the widespread extent of the vulnerabilities of many organisations – including, ironically, the UK spectrum regulator – to such instances. The wide spectral range of the frequencies affected, and the day of the week of the market crash, caused one commentator to dub the day 'Rainbow Wednesday' and the name seems to have stuck.

The UK government eventually stepped in and applied special powers and there was a brief return to command-and-control processes in those bands. Although this was an extreme example, there are many low-level instances of spectrum congestion having a significant impact on business efficiency and continuity.

At the same time, there are large amounts of spectrum that are not densely occupied. For example, large amounts of spectrum have been released following the analogue TV switch-off, although it is still not yet clear how this will be used.

## 12.4 The Age of Wireless Mundanity: 2017 to 2026

### 12.4.1 Introduction

As we approach the end of the third decade of the millennium, the old subject of 'wireless' has ceased to be a specific discipline in itself. The technological issues have been largely dealt with, and we take it utterly for granted that we can obtain connectivity between devices pretty much whenever and however we wish to. What continues to evolve, however, is the way in which we use and apply the technology for our needs. There is no particular dividing line between the domains of computing and communications: although 'convergence' went through many hype cycles, it now seems absurd that it once seemed sensible to interact with voice, entertainment, business and shopping services over different devices, interconnected via widely differing networks. Almost every financial transaction, every exchange of data between separated locations and every creative act now involves wireless communication at some or several stages.

### 12.4.2 Services and Applications

Wireless is now very much a standard part of a wide range of basic activities. After many false starts, domestic automation is now pretty standard. New light switches, locks, fridges, washing machines and the like have individual addresses and low-rate wireless transducers, allowing users to control and program interactions between these devices. This saves an enormous amount of domestic electricity and water, while also reducing wiring costs. Although the technology to enable this has been available for many years, it took longer than many had expected to reach the mass market.

Similarly, large-scale commercial buildings come with wireless infrastructure built-in as a standard part of the build specification. This is now as much a part of the standard 'fit-out' for buildings as standard power and ventilation previously was. As a result they gain in space efficiency, as well as providing essential services to the users of the building.

A whole host of security challenges were faced initially in the implementation of both domestic and commercial systems. Manufacturers were initially dismissive of the possible consequences, but then drive-by hackers began spelling out their names in lights on offices in major city centres. The associated publicity caused vendors to withdraw systems and rapidly agree a tight protocol for device security, leading to a good degree of resilience and reliability in current systems.

In the home, it is commonplace to be able to locate most possessions using your home network and to automatically route audio and video content according to the location of users and their preferences. Data backup of several terabytes per user (all of their lifetime's data) is achieved transparently and automatically across wide area wireless networks using cheap off-peak capacity.

As users move around the home, the embedded technology ensures that content and data is delivered to the right people at the right time. Voice recognition technology is finally the input device of choice, together with bio-emotive feedback: gestures in the air used to control activity, mostly replacing the keyboards, mice, remote controls and game controllers of the past.

### 12.4.3 Devices

The domestic environment in new houses now comes fully fitted with technology devices that are interconnected wirelessly. These include loudspeakers and projection displays, light switches and heating controls, microphones providing automated voice recognition and cameras monitoring detailed activity. They interact with biometrics, RFID tags and personal implants to deliver rich domestic applications.

### 12.4.4 Infrastructure

Network operators and service providers are very often separate entities now. Local authorities and commercial property owners own the housings for networks, including street furniture (lamp-posts, CCTV poles etc.) and rooftops. Major networks, spanning a wide range of technologies and users, are owned and operated by wholesale integrators, some of which evolved from the tier-one base station vendors of the past. Service providers then buy capacity on these networks wholesale and provide airtime and packages of differentiated services to end customers. In the main they are not responsible for the creation or provision of specific content, which is created by major media organisations and applications service providers. Little software is bought outright.

The networks themselves consist broadly of wideband radio heads, comprising relatively little intelligence beyond signal conditioning, but very fast analogue–digital–analogue conversion, significant storage (for when backhaul bandwidth is limited) and RF which spans from hundreds of megahertz to several gigahertz, with backhaul amongst heads provided by both fibre and fixed wireless across unused frequencies. Most of the real processing is provided in a tiered, distributed fashion. In this way both coverage and capacity may be routed to where it is needed very fast, with frequency, space and time allocations not preplanned at all. The conventional, inflexible 'radio planning' approaches of the past are irrelevant in this process.

Huge performance gain is available via these distributed base stations relative to the inflexible systems of the past, via 'macro MIMO' which forms effective 'bubbles' of capacity and coverage around each user, delivered via several nearby antennas which can contribute useful coverage or diversity. This allows them to realise immense trunking gains whereby a very wide bandwidth may be occupied for short periods and allows interference to be avoided as necessary via a mixture of ad-hoc and cooperative schemes. The network no longer has to be dimensioned for traffic peaks, since the capacity is moveable. A classic example of this is in cities, where formally the city centre network was under-utilised in the evenings, while those in the suburbs where lightly used in the daytime. Now the same processors and spectrum resources are applied for users distributed right across the city, routed via appropriate antennas according to the user locations. Given flexible spectrum allocation policies, it is usually best for systems to achieve maximum diversity to multipath interference, co-channel interference and other sources of interference by spreading signals over as wide a frequency range as is locally permissible.

The wholesale integrators have thus been able to realise a substantial return on their considerable investment, as radio heads are produced in enormous volumes, reducing their production and replacement cost, and allowing them to have a long lifetime as they no longer become obsolete when air interfaces are refined and enhanced.

### 12.4.5 Air Interfaces

Wireless air interfaces are now about as interesting as the Ethernet PHY was in 2007 – essential and ubiquitous but entirely undifferentiated. Many of the key patents involved have anyway expired now. The analogy with Ethernet is good, since Ethernet actually consists of a large and diverse family of physical interfaces; similarly many air interfaces now coexist on the same devices and infrastructures without competing in any real sense.

The endless battles between air interfaces (TDMA/CDMA, flavours of Wi-Fi, flavours of UWB etc.) are simply irrelevant – software-defined radios now select their wireless personality according to location and performance. Operators compete on services, cooperate on infrastructure and are largely uninterested in technology.

### 12.4.6 Spectrum

From around 2018, governments started to realise that the high-profile spectrum trading disasters were not the result of too much liberalisation, but rather too little. Although trading had been permitted in many bands, the trading was possible only within and amongst users of that band. It was not possible, for example, to trade an assignment in one band for several in others if that serviced the purposes better. Finally governments 'kicked down the walls' between the bands, and it became clear that, provided the right technology is used, spectrum taken as a whole is not scarce at all, but highly abundant. It was initially thought that exceptions should be made for safety-of-life applications such as aviation and emergency services, but in fact they gained more in terms of resilience and capacity than they lost from guaranteed access to a narrow set of bands, and they have retained their ability to create high-priority interrupts to other band users.

On the other hand, it has been widely recognised that international interests are now critically dependent on the efficient use of spectrum and can be very adversely affected in extreme cases by events such as the 'Rainbow Wednesday' crash of 2016 and the sunspot disruptions of 2010. As a result the principles of overall public ownership and control of spectrum are now enshrined in many national and intra-national constitutions. These principles are reflected in legislation such as the Spectrum Misuse Acts.

We now find that spectrum availability is rarely a constraint on services, and users now have access to as much data as they need at the rate and quality they need it, although sometimes at a significant price if not reserved well in advance. This has come about because of:

- Spectrally efficient air interfaces, and most of these gains came about during the Age of Wireless Similarity. These gains were just a few times the efficiencies that were available in 2007 – certainly less than an order of magnitude. The key techniques were a flexible, time-varying mixture of multiple carrier wideband CDMA, with MIMO and turbo coding closely matched to source and channel characteristics.
- Smaller cells, with flexible distributed spatial, spectral and temporal processing and resource allocation. Gains of typically one to two orders of magnitude compared with 2007 levels for macro-cells are routinely available.
- An ultimate limit on the data rate requirements of users. Although services are continuing to proliferate, the peak data rates consumed by users stabilised in around 2020 and networks are now able to supply this peak for all users. Once almost all users have access almost everywhere to average data rates of several hundred megabits per second in active periods, their senses are no longer able to make use of more data and higher data rates become irrelevant.

## 12.5 Conclusions and Summary

It can sometimes seem, working with wireless on a daily basis, that the future has already arrived. I wrote the majority of this chapter on the island of Gozo in the Mediterranean, where horse-drawn carts for small-scale agriculture are still a common sight. I worked on a mobile phone which supports seven air interfaces. The input device was a folding keyboard, working over BlueTooth. I researched the content and backed up intermediate versions over the air using GPRS. I continued to receive emails as and when they arrived, occasionally making calls by speaking the name of the person I wished to call. All of this would have seemed like science-fiction in the mid 1960s, but is easily available today. The remarkable thing is not that it works, but that it works easily: I expected coverage to be available and for these services to require zero configuration on my part – and that was indeed the case. This kind of ease of use will increasingly be taken utterly for granted.

Nevertheless, there is plenty of scope for wireless progress, and I hope I have explained the reasons and directions for these. In summary, my predictions for the future are as follows:

### 2007–2011: The age of wireless proliferation
- Data-hungry applications rise, producing substantial loading on existing wireless networks. Key amongst these are always-on email, instant messaging and mobile TV. Mobile video telephony fails to take off.
- Cellular networks provide increased depth of coverage for 3G to serve data applications via denser mobile networks, micro-cell networks on street furniture and some enhanced coverage.
- Wi-Fi systems begin to provide wireless PBX services in corporate environments, controlled by major operators.
- Many new small operators appear to exploit technologies such as WiMax in newly available spectrum. Many fail by not attracting sufficient customers away from large-scale networks. Some are successful in providing niche broadband connectivity services, but there is little take-up for mobile services.

- Domestic cellular base stations ("femto-cells") are extensively deployed and used to provide enhanced cellular services and to reduce customer churn.
- Device storage increases to the point where personalised TV channels can be created on the device from synchronised information streams, evolved from podcasting technologies.

### *2012–2016: The age of wireless similarity*
- Large data volumes are associated with machine-to-machine applications.
- High fuel prices drive the start of mass market in wireless building automation systems.
- Short-range high-data-rate applications for domestic video distribution rise.
- Many air interfaces are drawn together in 'meta' standards enabling flexible use of the best available scheme for a given environment and device/application/environment combination.
- Major consolidation between fixed and mobile operators occurs.
- Digital broadcasting: both fixed and mobile over dedicated networks comes to the fore.
- Wireless technologies and frequency bands are no longer closely linked.
- There is increasing network sharing and ownership by dedicated infrastructure companies and vendors.
- Spectrum hoarding and scarcity in particular bands forces increasing pressure to adopt flexible spectrum assignment procedures together with tight regulatory oversight of usage.
- Significant backhaul congestion drives increasing use of wireless backhaul.

### *2017–2026: The age of wireless mundanity*
- Wireless is an 'embedded' technology, taken for granted in most applications.
- Wireless building automation and networks in homes and public buildings are widespread.
- Service providers, network operators and content providers become very separate. Infrastructure sharing is common.
- There is flexible allocation of services to appropriate air interface, frequency band and group of antennas over programmable, wideband infrastructure with a very large number of service points and extensive signal processing to select best time/space/frequency combination for each service at a given time instant.
- There is spectrum abundance relative to data requirements.

I am confident that many of these predictions will be proven true and I have equal confidence that I will be utterly wrong in some important respects. Indeed, in several cases I have included predictions that may seem extreme in order to provoke discussion and to illustrate some important points. I look forward to discovering which predictions become realities over the coming years. In the meantime, I welcome hearing of your own opinions and your comments on mine at my website and blog: www.simonsaunders.com.

## Reference

[1] David H Hathaway and Robert M Wilson, 'What the sunspot record tells us about space climate', submitted to *Solar Physics*, August 2004, NASA/Marshall Space Flight Centre/NSSTC, Huntsville, AL 35812.

# Biography

### *Professor Simon Saunders, PhD, CEng, FIET*

Simon Saunders has specialised in wireless communications throughout his career, with particular interests in radio wave propagation prediction in urban and indoor environments, smart antennas, and mobile systems engineering. He has pursued these interests in both industry and academia. As Chief Technology Officer for the Red-M Group, Simon developed and delivered services and products via a best-practice approach which he originated, known as 'Total Airspace Management'. He is also a Visiting Professor at the University of Surrey. Simon has acted as a consultant to a wide range of companies, including BAA, BBC, Mitsubishi, British Land, O$_2$, Ofcom, BT, NTL and many others.

Professor Saunders has invented more than fifteen patented technologies and has written more than 130 publications in learned journals and international conferences. He has also written a successful book, *Antennas and Propagation for Wireless Communication Systems*, which will appear in a second edition shortly.

He obtained BSc (Hons) and PhD degrees from Brunel University in 1988 and 1991 respectively, receiving four separate awards for academic achievement in that time. He worked on microwave oscillator structures as a Research Fellow for Brunel in 1991. In 1993–94, he proposed a novel parallel computing approach to the solution of electromagnetic problems and acted as consultant to the resulting project, which led to the creation of an electromagnetic simulation package, used as a CAD and teaching tool.

For several years Professor Saunders worked in an advanced development role for Philips Telecom, including the standardisation of the TETRA air interface for digital private mobile radio. He led a team designing technology for a novel two-way paging system. Simon also spent one year working with the Ascom in Switzerland, defining product technologies and system design solutions for the first generation of TETRA handsets and base stations.

Professor Saunders worked for Motorola in their GSM Research Group in 1995, designing air interface and system planning enhancements leading to increased capacity second-generation GSM systems. This included several patented advances in antenna, modem and cellular architectures.

# 13

# Mobile Cellular Radio Technology Disruption

Stephen Temple CBE

## 13.1 Extrapolating from the Past 25 Years of Public Mobile Radio

The first two generations of cellular radio have shown just how disruptive a new generation of cellular radio can be. Prior to 1982, public mobile radio systems operated in the VHF frequency bands, they had limited capacity and no handover of telephone calls across coverage areas. The phone needed the car's boot to carry the equipment and the car battery could be drained by a long call. The limited capacity led naturally to rationing by price. It was a service only the very rich could afford.

First-generation cellular radio systems came in during the early 1980s. The technology breakthrough was the cellular architecture that significantly increased the capacity of the networks. Customer convenience of a car phone service was enhanced with handover of calls as customers crossed cell boundaries. Most of all, the extra capacity transformed the economics of the business. The service was brought within the reach of the smallest businesses. Motorola introduced the hand portable mobile phone (the iconic 'brick') to the world. For the cellular radio industry it became big business. In a relatively short space of time the pre-existing VHF public mobile radio systems became obsolete and were completely swept away. The business largely remained a telephone service.

The technology disruption that drove the second-generation cellular radio networks was 'digitalisation'. Governments released more radio spectrum in 1991. It was tied to the introduction of the new digital technology. The extra spectrum allowed new market entry. It was a technology disruption given an added impetus by the disruption of a new competitive market structure. The digital data capability provided the basis for a new universal messaging service to develop (SMS). It was a highly successful model that drove cellular radio into a mass consumer product. The mass scale effects reinforced the success story. The competitive

*Wireless Communications: The Future*   William Webb
© 2007 John Wiley & Sons, Ltd

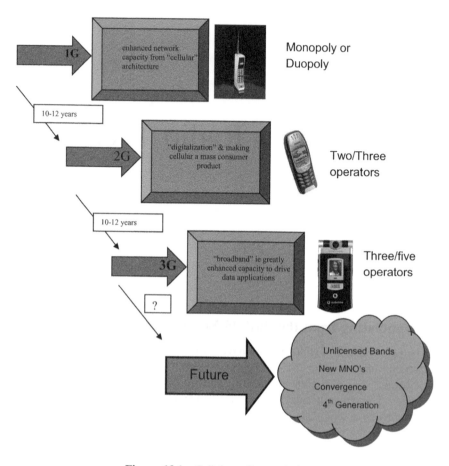

**Figure 13.1**   Cellular radio revolutions.

game was to offer more coverage. Profit in the urban areas was high and this allowed GSM networks to be pushed into a substantial fraction of unprofitable habituated rural areas. For most people the personal mobile phone became the third most important thing to have in the pocket after the house keys and the wallet. The analogue cellular radio networks became obsolete and were also swept away. The time between the initial introduction of the first-generation analogue cellular networks and initial introduction of the second-generation digital networks was 10–12 years.

This brings us around to the third-generation cellular radio networks. The technology disruption was broadband data speeds. The promise was the introduction of new multimedia services. The time between the initial introduction of the second-generation digital cellular networks and initial introduction of the third-generation wideband networks was again 10–12 years.

With three points already plotted on a chart such as Figure 13.1, it seems simple to suggest that in 10–12 years from the introduction of 3G-technology (say in 2014) another disruptive cellular radio technology will arrive in the market. The second-generation digital networks (GSM) will have been just swept away and the writing will be on the wall for the ultimate demise of the third-generation technology. A number of leading research laboratories are already suggesting that the fourth-generation disruption will be an end-to-end IP network running at data rates between 100 and 1000 Mbits/s (see Table 13.1).

**Table 13.1**   Vision for 5–10 years out based upon simple extrapolation.

- Fourth-generation cellular radio technology introduced in 2014.
- IP end-to-end.
- Customer data rates> 100 Mbits/s.
- GSM networks closed down.

Could this study be that easy to write? Almost certainly not! The clouds of convergence are starting to fog the scene. IP services such as VoIP are already beginning to run across 3G networks. We are not even at the half-way mark of our 10–12 years of the supposed heyday of 3G multimedia services, and TV is starting to run over new networks that are definitely not 3G networks. MP-3 players have swept in to provide music on the move. WiFi hotspots are multiplying in some dense urban areas like a rash. Suppliers are making stunning claims about WiMax.

We need to take a much closer look at what is going on. A good starting point is to reflect on how easy or difficult it is to make disruptive changes to communications network infrastructures.

## 13.2 The Law of Large Network Momentum

It is very difficult to get a new network technology off the ground. Nobody wants to buy a terminal until the network is widely deployed. There is no money in widely deploying a network until there are a lot of terminals in the market to drive traffic through the network. Then for the terminals themselves they are expensive to develop and only become affordable if this cost can be spread over a huge number. But a large number will be sold only if the price is low. This can lead to unvirtuous circles. Digital audio broadcasting in the UK is a good example of an unvirtuous circle that lasted for nearly a decade before DAB receivers finally crept past the tipping point.

However, once a tipping point is reached the conditions invert. Large-scale economies drive down terminal prices that attract even more customers that drive down terminal prices. This fuels networks to be driven out further to capture more potential customers.

Essentially it is a bi-stable market where networks are either in an unvirtuous circle or a virtuous one. Very large forces (including financial ones) are needed to flip from the bottom to the top. Once on top there can be a self-sustaining momentum. A parallel is one of Newton's laws of motion that says that a body (read here a successful network) will continue in a state of uniform motion until compelled by an external force to change that motion. Two great examples of this have been the analogue TV service and the analogue plain old fixed wire-line telephone. Huge sunk cost for the network investments and scale economies of terminals sustain their position unless and until they face a large disruptive force (or forces). In the case of the analogue TV service we see the external force of government acting to finally apply the brakes.

The approach we will take is to look at the 3G world 5–10 years from now in the absence of any disruptive forces, round up the usual suspects of disruption and take a speculative look at how things may then play out under the influence of 'blindside' disruptive forces.

## 13.3  Third-generation W-CDMA Future

One of the basic tenets of cellular radio networks for the first two generations was that they were 'horizontal' in so far as the market was concerned. They served both the consumer and business markets, providing a common platform for all applications. This basic tenet of the cellular radio network is likely to break down during the life of the third-generation networks. The reason is that cellular radio operators are looking for growth opportunities in adjacent markets and in particular to support entertainment services in the consumer market and mobilising IT applications in the enterprise market. The pressure this applies to the 3G networks is shown in purely illustrative form in Figure 13.2.

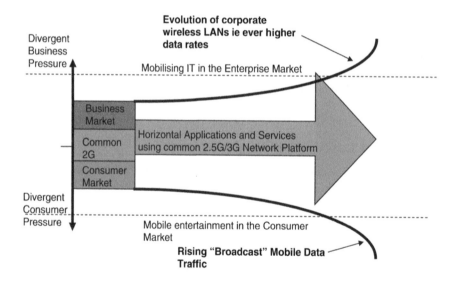

**Figure 13.2**  Challenging the basic tenet of the purely horizontal network.

Time is shown on the horizontal axis and the vertical axis is a representation of the extent of departure needed from the basic system design to address adjacent market needs. The dotted horizontal lines illustrate the fundamental boundary constraints of the 3G technology; i.e. the limits to which it could be stretched and remain backwards compatible.

This pressure has been compounded by the 'under-performance' of the first release of the 3G technology (called Release 99). At the time of the 3G auctions cellular radio operators were led to believe by the suppliers, media and governments anxious to get a good price for the spectrum that the 3G networks would deliver 2 Mbits/s data speeds to customers. What they failed to mention was that that would be possible only for one customer in a cell and that customer would have to be hugging the radio mast or pretty close to it. The performance in a typical cell for the Release 99 is around 500 kbits/s and this gross data rate has to be contended for by the number of customer simultaneously trying to receive data. This left the mobile network operators on the one hand unable to meet, for any length of time, the full spectrum of needs of their marketing ambitions, and on the other hand vulnerable to disruptive entry of specialist networks coming out of the consumer space and enterprise space, as illustrated in Figure 13.3.

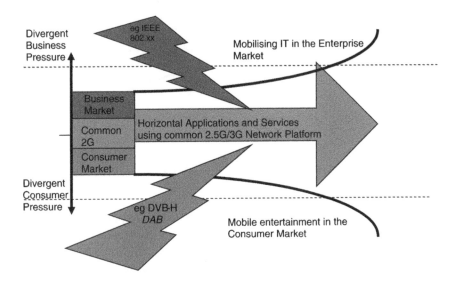

**Figure 13.3**   Competitive threat to the purely horizontal network.

The result of this imbalance of market ambitions, network performance and competitive threat will to be drive the 3G technology on a track of continuous improvement, and this is illustrated in Figure 13.4.

HSDPA (high-speed down-path packet access) is an upgrade for the data rates from the network to the handset that switches to more efficient modulation methods and codes for those users closer to the base station. In fact HSDPA is a whole roadmap of upgrades described as Cat 12 (up to 1.8 Mbits/s), Cat 6 (up to 3.6 Mbits/s) and, just to confuse us, Cat 8 (7.2 Mbits/s) and Cat 10 (up to 10.8 Mbits/s). In addition to supporting higher peak rates, the total capacity of each cell is greatly increased. With advanced terminals this capacity can be as much as six times. At the same time, the maximum speed from handset to network will increase from 64 kbits/s to 384 kbits/s, although not all users will be able to achieve the maximum rate.

The HSDPA upgrade is inevitable. It gets the cellular radio operators back to where they thought they were when they entered the 3G auctions. In fact it is likely that any new operator putting in a 3G network for the first time will launch it from the outset with HSDPA carriers.

Further speeding up of the data rates from the handset to the network will be accomplished by an upgrade technology called HSUPA (high-speed up-path packet access). This will double the capacity of the network, increase the availability of 384 kbits/s and enable users over more than half the cell area to achieve speeds of 1–2 Mbits/s. This will

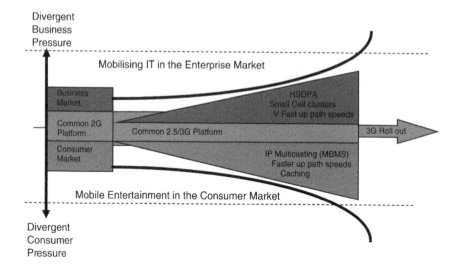

**Figure 13.4**   The 3G technology is on a path of continuous upgrades.

be attractive both in the business market (up-loading PowerPoint presentations) and in the consumer market (for up-loading photos from cameras with rising pixel resolution).

In dense urban areas we are likely to see a growing density of 3G base stations. This will be driven by the desire to deliver these high speeds over greater areas (in competition with WiFi hotspots but trading lower speeds with much greater mobility) and to improve indoor coverage.

In spite of all this effort there will be a limit to the extent to which the 3G networks can be stretched. There will be an inevitable competitive fight at the boundary. Mobile TV is a good example of this in the consumer market, which we will come to later.

## 13.4 Fourth-generation Technology

There was a time where the marketing departments of all the technology contenders that had lost the race to become the third-generation cellular radio standard re-badged their technologies as 'fourth generation'. This became so counter-productive that the world's cellular radio community coined the term 'beyond 3G'.

Already in my analysis I have shown that there will be specialist mobile networks *sitting alongside* (as opposed to *beyond*) 3G that will complement and compete with the 3G technology. A starting premise might be that no technology goes on forever and therefore something must emerge in the future that will eventually displace even the most stretched of 3G technologies.

The first challenge to this premise is the emergence of a 'combination of technologies' of 3G plus WiFi plus (say) DVB-H. This could close down the capability gap of the current 3G technology. Certainly modern silicon technology advances look capable of shrinking down the chips of three receivers and two transmitters to little more that a single transmitter and receiver. Once the three networks are widely deployed and the investments become a sunk cost and a huge installed base of mobile handsets exists in the market that can work with any of the three technologies, then the bar is well and truly raised for a fourth-generation technology. What can it possibly do much better than our three-way combination?

The consensus in the R&D community is that a true fourth-generation network will be end-to-end IP and run at data speeds between 100 and 1000 Mbits/s.

The transition to an 'all IP' network could occur with the 3G technology. The significance of the upgrade to HSDPA is that it will work efficiently with an IP core backbone network. All that is then needed is a VoIP client on 3G handsets with sufficient processing performance and the way is open to gradually replace the conventional space-time switches of current mobile telephone exchanges in the network by IP routing servers. This then offers IP voice, data and video services end to end. WiFi is already a purely IP medium. All that would remain out of step would be dedicated mobile TV technologies such as DVB-H. The serious point of this is that, whilst a fourth generation cellular radio technology may be 'all IP', that is a long way from fourth generation being the enabler for an 'all IP' cellular radio revolution. In other words it will not be a disruptive reason for anybody to pour investment into it to get it off the ground. This leaves the issue of speed.

Since the history of radio began there has been a race between radio and wire-line means of communications to provide higher speed data connections – and line systems have always won. Then 'mobile' radio systems have a huge handicap over line-of-sight radio systems in not putting customers to the inconvenience of having to point their mobile phones in a particular direction to make telephone calls. This loses valuable antenna gain on the mobile terminal. Finally the handset is both power limited and RF safety limited. That said it has been and remains a challenge for mobile radio engineers to research new boundaries to push up cellular radio data speeds. This is likely to come from at least two different sets of industries. The first is the cellular radio suppliers. The second is the IT industries. Both are seeking to chase the wire-line speed for which today's benchmark is gigabit-Ethernet. To support such high data rates requires very wide bandwidth and the space to accommodate this is only likely to be found above 2–3 GHz (see Figure 13.5).

Achieving ultra-high data speeds has quite profound implications for the ubiquity of coverage of a cellular radio infrastructure. This is illustrated in Figure 13.6. This leads us to argue on economic grounds that such a fourth-generation cellular radio technology will be targeted at dense urban areas only and very unlikely to ever get rolled out to less populated areas.

In fact already this is already becoming manifest with 3G technology. The traffic demographics show that the 20% of mobile operators' current GSM base stations are carrying just 2% of their total traffic. The case for rolling out even 3G services to this last 20% of base stations is at best flimsy. Worse, GSM rural coverage has been designed for outdoor coverage. To not only translate this into indoor coverage but also provide very high data rates would require a huge increase in the density of base stations. Rational cellular radio operators are going to baulk at the idea of having to find the additional rural base stations needed for high data rates indoors for 3G coverage – even if the 3G technology is rolled out in re-farmed 900 MHz spectrum.

These basic economic fundamentals may well drive cellular radio operators to sink their competitive differences and share 3G infrastructure beyond the 80% of the population coverage, and this may drive the 3G infrastructures out to the 90–95% of the population levels leaving GSM dealing with the less dense rural areas.

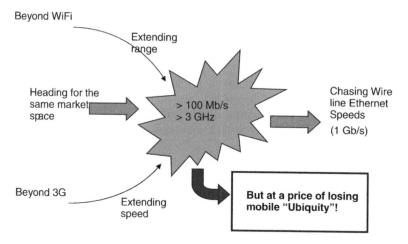

**Figure 13.5**   One likely technology track for beyond 3G technologies.

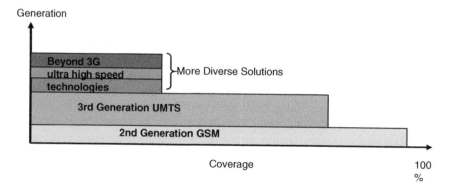

**Figure 13.6**   Diminishing coverage for higher speed mobile networks.

Based purely on economic factors, the most likely cellular radio infrastructure trend is newer higher performance network technologies getting rolled out to ever diminishing geographical areas. Dense urban areas will be massively over-served, suburban areas less so and rural areas stuck in a time warp.

## 13.5  Where does this Leave the Switch-off of GSM?

The transition from second-generation technology to third-generation cellular radio technology is likely to be much more complex than the transition from first- to second-generation technology, for the following reasons.

## Rural coverage

Cellular radio operators will want to hold on to their rural coverage as a service differentiation in a market where urban areas will become over-served by competition. Many are likely to leave their GSM network running for this purpose.

## Roaming traffic

New operators in emerging countries are not dashing to install the latest 3G technology in quite the same way they did in the analogue cellular radio days when GSM first emerged. The biggest growth markets are the emerging countries, and most of this expansion is running on GSM technology. Therefore, five years from now there will be significant number of people coming into Europe with GSM-only handsets. The income from this roaming traffic is more than enough to pay the running cost of a GSM network.

## Rump of GSM customers in Europe

Studies show that even after another five years of marketing 3G services in Europe there is likely to remain 15–20% of customers that simply are not attracted to 3G. It will cost cellular radio operators a lot of money to force-migrate this rump of GSM customers over to 3G. If the GSM networks are left running for roaming and rural coverage, it makes commercial sense to leave the new technology 'refuseniks' with their GSM service.

## Terminal parity

There have been serious efforts by suppliers, led by a GSMA initiative, to produce the $25 GSM mobile phone. This has already been achieved and even lower prices are emerging. On the other hand, the intellectual property rights for the 3G technology were left in a bit of a mess by the standards body. As a result, 3G terminal prices have not decayed on the curve that might have been expected for the volume of terminals being bought. If we project what the lowest price terminals are likely to be five years from now, we come to the surprising result that the lowest price 3G terminal may cost more than three times the lowest priced GSM terminal. This is a huge gap for a competitive market to ignore, particular at the no-frills light user end of the market. The very low cost of GSM technology is likely fuel a tidal wave of new applications for GSM involving voice and relatively low data rate services that will comprise:

- *Low-end telephone users.* Around 20% of the market wants the security and convenience of a mobile telephone but the usage is relatively low (e.g. fewer than ten telephone calls a month). GSM looks likely to dominate this light user segment five years out from now.
- *Machine-to-machine.* Consultants have been predicting for years that a machine-to-machine mobile data market is about to explode. It has not yet and one reason for this may well be the cost of a GPRS data module relative to the cost of the machine-to-machine mobile application. However, if a full GSM handset will cost less than $25 it is more than likely that a GSM data-only module will be under half this price. If the GSM modem is embedded within mass-produced products, which already contain electronics, the added cost could well be as little as $5. This sort of price point could be the trigger for the long predicted explosion in the machine-to-machine mobile data market, and GSM networks are likely to be the main beneficiary.

In other words, GSM networks will still be around ten years from now. Whilst individual cellular radio operators may have consolidated their GSM assets, GSM service will still exist over the very length and breadth of Europe. In fact GSM may well have become the mediation layer for a variety of 'network technology sandwiches'.

## 13.6 The 3G Cellular Radio Network Landscape Ten Years from now

This brings us to the stage of drawing together the results of the foregoing analysis into a projection of the European cellular radio infrastructure world ten years from now:

- The performance of 3G networks will be greatly superior to today's 3G networks and offer 'just good enough' data rates for wide area mobile applications to the small screen.
- The 3G networks will be the workhorse for wide area mobile across Europe with coverage gradually creeping out to around 90–95% of the population once a 900 MHz version becomes available.

But this 3G world will find itself hemmed in on several fronts:

- Mobile TV networks (e.g. DVB-H) will sit alongside the cellular radio networks in both a competitive and complementary form.
- WiFi will be successful particularly in the enterprise world with the total convergence of voice and data in the office.
- Sitting under this will be a GSM network that will have experienced a rebirth for light data applications. Any box in any industrial or consumer segment, whether static or mobile, in any location that has any useful data will have a GSM chipset to get that data back to where it is needed and similarly for controls in the other direction.
- Radio spectrum will be more readily available than it has ever been. This will encourage a plethora of new technologies to be tried out in the market, but the sophistication of the terminal industry in putting combination of technologies (e.g. 3G/GSM/WiFi, GSM/DVB-H, 3G/WiFi) will make the area of unsatisfied customer need very small and most of these new technology hopefuls will bump along the bottom of the curve.

This has provided a possible scenario in the absence of disruptive forces (see Table 13.2).

**Table 13.2** Vision 5–10 years out shaped by the law of network momentum.

- 3G (in its various stretched forms will become the main wide-area data workhorse.
- WiFi will dominate the enterprise space. WiFi will also be widely used in the home.
- GSM will re-invent itself for low-end light data use (e.g. machine-to-machine).
- This technology triumvirate will only be challenged where there are customer needs that these technologies cannot be stretched to meet:
- Mobile TV networks will emerge to offer more immersive TV on the larger mobile screens.
- Very high-speed mobile networks will emerge to gradually challenge WiFi for enterprise networks.

We now come on to the first of the likely disruptions and that is 'convergence' of technologies, industries and markets.

## 13.7 Convergence as a Disruptive Force

In the very early 1980s one could have characterised the communications service space by four circles barely touching:

- *Fixed telecommunications.* Inside were the dominant PTTs. The PTT was the telecommunications operator, regulator and government all rolled into the one dominant entity.
- *Mobile operators.* This comprised, in Europe, largely duopolies. One side of the duopoly was the mobile arm of the PTT that was spun out and operated quite independently of the fixed operation. The other side of the duopoly was the new GSM challenger.
- *Broadcasting industry.* Well entrenched were the free-to-air broadcasters (public and private). In the Benelux countries extensive cable TV networks existed. The first of the satellite pay TV operator just appeared.
- *IT industry.* In 1982, IBM introduced the PC that was a prerequisite to the eventual emergence of the Internet in the consumer market.

If we now roll forward 25 years we find the Internet at the very epicentre of service 'convergence'. The flat rate 'all you can eat' pricing model of the broadband Internet is making 'bandwidth 'free'. It is a disruptive business model sitting on top of a disruptive technology (IP protocols as a universal service glue). The knock-on effects in the communications industry will be felt over the next 5–10 years:

- *Fixed operators* have aggressively seized the new opportunity of the broadband Internet and have sown the seeds of the destruction of their telephone voice revenues through VoIP over the Internet in doing so. Their voice revenues have also been under attack from the mobile operators. They are reacting by pulling back in house their mobile operator arms and developing triple and quad service plays.
- *Mobile operators* that have no fixed infrastructure have been looking to deepen their attack on the fixed voice market as well getting into the entertainment space, including video, music and TV. Their new 3G mobile phones are becoming more and more like PCs and potentially an integral part of the Internet. Advertising on the mobile phone will create both complementary and competitive relationships with the fixed Internet players.
- *Broadcasters* have remained a relatively consolidated industry but see the Internet players and mobile phone companies lapping at the edge of their territory. The Internet coupled with storage technology is a huge potential disruption in allowing viewers to break free from the grip of the 'programme schedulers'.
- *IT industry* (defined here as including the Internet players) dust has settled after the dot.com meltdown and a handful of giants have emerged to dominate the landscape, each with a business model having quasi-monopoly characteristics. Google has an almost unassailable position in search-related advertising, eBay in on-line auctions; and Microsoft's 'Windows' remains powerful.

We look in turn at a few of the more significant zones of convergence for the cellular radio industry

### 13.7.1 Convergence: Mobile and Broadcasting

Mobile TV has been launched across 3G networks by streaming the TV channels. Each customer is sent his or her own data stream even though watching the identical TV programme to another customer just a short distance away. This is a very data-hungry means of distributing television. If each data stream is 128 kbits/s then the gross data capacity needed in a cell is $n$ times 128 kbits/s. Since the total cell capacity of Release 99 is only around 500 kbits/s, it is obvious that this means of distributing television in cells having many hundred customers in busy times is not sustainable. The upgrade to HSDPA brings extra data capacity and buys the cellular operator time. However, it still remains very inefficient to distribute broadcast television as separate data streams for identical TV content. It means adding network capacity (more base station sites) as the number of customers increase and this is expensive.

In a broadcasting architecture only one stream is sent per TV programme and all customers wanting to view that TV programme get the same data stream. This removes any limit on the number of customers watching that TV channel.

There is an upgrade of 3G called MBMS that allows one data stream to be sent to multiple viewers all demanding to watch the same TV channel. The statistics of how many viewers watch the most popular TV channels shows that typically six TV channels are needed to lift 70% of the concurrent data peak off the HSDPA network. Since MBMS provides for a 1 Mbits/s gross data rate, then if we need to lift this amount of broadcast traffic off the HSDPA network we have to limit the data speed to 150 kbits/s per TV channel. Fortunately this data rate is enough to give excellent picture quality on small mobile handset screens (1.8–2.0 inches).

For mobile phones with larger screens the picture does not look as good. Blocking effects can occur for fast-moving picture content. Thus the boundary line between what 3G technology can do is set effectively by the screen size. If customers want to watch their mobile TV on larger screens (say up to 3.5 inches) either the TV channel choice has to be limited or a larger broadcasting data bearer is needed. This is providing the opening for dedicated broadcasting technologies. The mobile broadcasting technology getting the most traction in Europe is DVB-H. There are other contenders.

The most economic deployment of DVB-H provides a gross data rate of 3.74 Mbits/s (QPSK) or a more expensive deployment (16QAM) needing twice the number of base stations to provide 7.46 Mbits/s. The economic deployment could provide for nine TV channels at 384 kbits/s that might be divided up in a two-operator infrastructure sharing arrangement of five shared TV channels (e.g. free-to-air) plus, say, two operator specific pay channels. TV channels at a data rate of 384 kbits/s would provide good quality picture on the larger screen sizes.

The trend for these more specific networks to emerge over the coming decade is inevitable in the consumer space. What is not inevitable is their extent, just how much of the market for their intended applications they capture and whether they hit a point of inflexion that lifts them from being present on a limited range of high-end handsets to become ubiquitous on almost all handsets.

The answer to all these questions will depend on the interest consumers show in paying for mobile TV consumption. If a huge market reveals itself we may well see even more advanced mobile TV technologies emerge in Europe that are likely to feature a satellite transmission 10 000 or so terrestrial repeater stations per country to ensure indoor coverage. There are two distinct uncertainties in the future mobile TV market:

- The first is whether customer interest in watching scheduled TV on tiny mobile screens will be sustained beyond the novelty phase. The obvious action here is to make the screens larger to allow the viewing experience to be more immersive. But this runs counter to the trend of mobile phones becoming smaller (e.g. Motorola's style leadership Razr and krzr phones). A probable outcome is that the handset market will segment between larger screen handsets that are pitched to those who want to watch mobile TV (and play games) and those who want something smaller and lighter to put in their pockets but might be interested in 'snacking' on TV to fill-in time when waiting for a train or in the car-wash.

- The second volatility is between those who want to watch scheduled free-to-air broadcasting and those who record their favourite programmes and watch them later at their own convenience.

The iPod is already successful and is expanding from music to videos. If we add the ease of moving content from the PC to the mobile phone to the fact that storage capacity in mobile phones is set to go up in leaps and bounds, we have the ingredients in place for a video viewing experience that may well compete with mobile TV for the attention of consumers. Where playing video straight off the storage of the mobile phone memory can outgun mobile TV is in providing truly universal mobility. It will work in buildings, the most remote rural setting, in aeroplanes in tunnels . . . everywhere. 'Mobility' has demonstrated its strength as a market driver in its own right.

We are likely to see a three-way split in the market between those having no interest in using their mobile phones to watch any sort of video, those watching varying amounts of live mobile TV and those pod-casting their video programmes (and only wanting live TV for the odd breaking news flash). Sitting behind this customer behaviour will then be two sorts of mobile TV network. One will be dedicated mobile TV networks (e.g. DVB-H) that are deployed but not geographically universal (i.e. mainly urban and suburban coverage) serving consumers carrying the larger screens. The second will be 'MBMS over 3G' networks to pump video programmes to 3G handset storage (mostly overnight) and deliver occasional news flashes during the day.

### 13.7.2 Convergence: Internet and Telephone Services

VoIP has already had a big impact on the fixed telephone service. People tend to focus on the technology. But the disruption has largely come from the common method of pricing for broadband Internet access. The customer pays an access fee to the huge open Internet cloud and usage is free. This has allowed third-party companies to run a telephone service on the back of the 'free data capacity'. Why has this not happened to the same extent in the mobile world? The most visible reason for this is that data on mobile networks is charged on a usage basis rather than a flat rate as we see on the fixed broadband network. This makes VoIP over a 3G data connection just as expensive as a normal mobile voice call.

It is inevitable in a competitive market for there tol be an over-supply of network data capacity and particularly in dense urban areas. At some point an over-capacity of data will drive the competitive market towards flat-rate data pricing and this will meet one of the two preconditions for third-party (parasite) VoIP to hit the cellular radio market. There is an inevitability that this will occur in most of the European cellular radio market in due course.

'The other pre-condition is sufficient 3G mobiles with the capability (including processing power) to support Voip. This is a less visible barrier and yet is more significant. It is why the media analysts predicting the immanent collaspe of the mobile market under the assault of Voip have got it so wrong. The huge installed base of mobile phones unable to support Voip is effectively shielding the cellular radio market from the full force of the potential Voip disruption. The real measure of vulnerability is the percentage of the entire installed base of mobile handsets that Voip clients can very simply self-installed. On this measure the impact of Voip is at least 3 year away and depending upon how the mobile operators respond could be as much as 5–7 years off.'

The cellular radio operators do not need to adopt VoIP technology themselves to simply reduce the price of their calls. The logical response of the cellular radio operators is not to give away huge flat-rate data bundles without also giving large voice minute bundles. This will take much of the steam out of VoIP over cellular radio. Another response could be for mobile operators' flat-rate data to offer generous data caps in the network to mobile handset direction (to meet all of the download market need) but limit the data cap on the mobile handset to network direction to a few hundred megabytes. This will hit heavy VoIP use that is symmetric in its data demand compared to downloads that are highly asymmetric.

VoIP over 3G cellular phones is an example where many of the pundits have correctly forecast something will happen but have misjudged how long it will take.

### 13.7.3 Convergence and the 'Battle for the Home'

The biggest battleground for 'convergence' will be 'the home'. Today most large global companies have grand designs to win 'the battle for the home'. Their starting point is Figure 13.7 and the epicentre of convergence is the broadband Internet. The broadband Internet is worth a closer scrutiny.

The Internet appears a turbulent space with time to market for new innovations fiercely short. However, sitting under this is a huge inertia of the installed base. Applications tend to be written around where the bulk of the market is in terms of access speed. Today this is probably around 512 kbits/s. At one extreme we still have dial-up customers content with their 64 kbits/s and at the other extreme those with connections of up to 10 Mbits/s. We also have extremes of customers in terms of the data they ship per month. At one extreme we have a few customers shipping a whopping 60–80 Gbytes per month (DVD peer-to-peer traffic) and at the other extreme people who simple use the Internet a few times per week to check if they have any personal emails.

Some data measured by a large European ISP on their broadband customer base in 2005 showed that 50% of their customers were shipping less than 600 Mbytes per month. Figure 13.8 illustrates the sort of activities customers are likely to be doing that results in the values of data being shipped each month.

Another way of looking at this customer usage is that it reflects a divide between customers for whom the broadband Internet has become central to their lives and those where it is only an adjunct.

We must not forget a large percentage of customers who have no connection to the Internet. Analysts are projecting that, five years from now, the average European penetration of the broadband Internet will be around 60%. (This average masks a spread of between 40–80% between different countries.) In other words, 40% of European consumers will not be connected to the Internet (but most of these are almost certain to have a mobile phone).

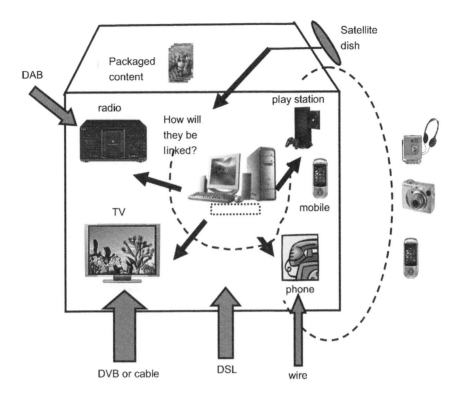

**Figure 13.7** Converging industries in a converging home.

These statistics allow us to characterise European homes in five years as belonging to one of three groups:

- broadband heavy-user homes
- broadband light-user homes
- homes with no Internet access.

The broadband heavy and light user homes will be the primary target for 'triple' and 'quad' plays. There is sometimes confusion over the terms 'quad play', convergent offerings and bundled offers. It is useful to re-define the possibilities as shown in Figure 13.9.

Simply offering customers several services (telephone, TV, mobile phone and Internet access) that have no technical or functional or service linkage with each other I have called 'bill convergence'. The only thing the four services have in common is the single bill from the same provider. That of itself is not a significant disruptive force. Who cares if the bills come in one envelope or three? More telling is where the provider drives the package by giving away one of the services below cost.

The next level of convergence I have called 'access convergence'. This is typically where other functions are integrated into the DSL access modem. It is a box-in-the-home play and starts to take on the attributes of 'the home hub'. Softbank has played the convergence of the telephone (using VoIP technology) and Internet access very successfully in Japan.

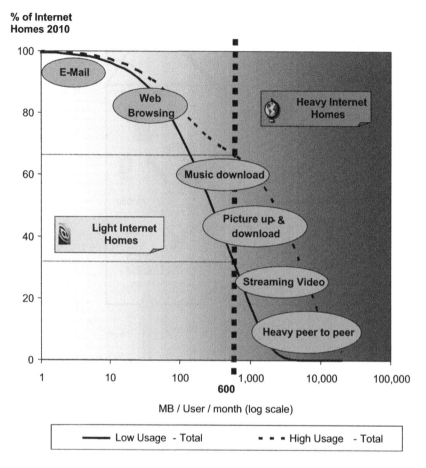

**Figure 13.8** Relationship between data usage and applications.

Free in France has added TV to the mix. Ease of access and distribution around the home is the key. Customer loyalty is the prize for getting it right and this improves business performance by reducing customer churn.

The third level of convergence is the most complex but likely to prove the most compelling to a much large number of consumers where it has been got right, and that is 'service convergence'.

### Broadband Heavy-user Homes

In the broadband heavy-user home the PC is already the de-facto 'home hub'. Almost certainly there is already WiFi connectivity extending the DSL access connection to multiple PCs in different rooms in the home. The PC hard drive is being used as the master store for a music collection and family photo albums. This is providing the initial connection to the mobile world as photos in digital cameras are downloaded to the PC, sorted and filed. MP-3 players are uploaded with the next bit of the music collection to be listened to on the move.

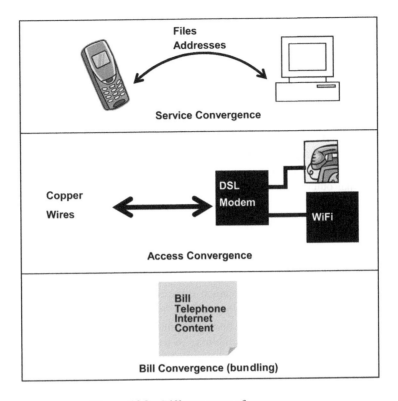

**Figure 13.9** Different types of convergence.

These will be the first homes to embrace WiFi-enabled mobiles to allow VoIP calls over the broadband Internet to be made at home using a mobile phone. The uploading of photos from the camera phone and downloading of music to the mobile phone is likely to be done more and more via a WiFi connection.

The three missing pieces in the majority of broadband heavy homes today are the full integration of the mobile phone with the PC (since the mobile is becoming a PC on the move), full interchange-ability of video content between the PC and TV sets and PC centric home security and environmental control. All this will come over the next 5–10 years with some content type constraints (certainly over the next five years), on moving content from the TV to the PC and vice versa for those homes that embrace high-definition TV.

Cheap low-cost bulk storage (e.g. hard disk drives) will see very high capacity storage creeping into many consumer electronic and other devices, leading to a demand to move data via some sort of home network between the various storage caches.

A big uncertainty for the broadband heavy homes is where Internet access speed will eventually plateau out. The most data demanding application is video, but there has been a huge advance in video compression technology. Here we have a technology trend and counter-trend. The other two forces in contradiction will be customer pull (happy with 512 kbits/s to 1 Mbits/s) and supplier push trying to drive up demand towards 10 Mbits/s.

At the moment there is a slightly deceptive game being played by some broadband access providers in appearing to offer ever higher DSL access speeds when, in fact, all that they are doing is lifting the speed caps set in the DSL modems. It is a bit like lifting the speed limit on a congested motorway. The actual speeds customers get are far lower when most customers want to use the service. They are lifting the speed cap without investing in lower 'contention ratios'.

That said, the Internet, with the necessary investment, is capable of running at much higher speeds. Some of that investment is in storage to cache content nearer to the edge of the network. Some of it is getting fibre optic cable closer to the customers and eventually all the way to the customer's home.

The most rational outcome will be small islands of very high-speed access connections in dense urban conurbations (e.g. flats) and this could be the starting place for the next generation of WiFi running ten times faster than today's WiFi.

## Broadband Light-user Homes

The broadband light user home will be much more diverse in terms of the technology in use. The cellular radio operators have their chance to win a sizable slice of these homes once they have cracked the challenge of linking the mobile phone and home PC worlds with the right combination of functionality, simplicity of use and price points.

Two trends are running in favour of the cellular radio operators' success in the broadband light user home:

- Broadcasters over traditional distribution channels (satellite, cable and terrestrial) and their supply industries will not give ground to the broadband Internet without a fight. The emergence of high-definition TV in Europe is one example of this. Another is the success of digital terrestrial TV in some countries. The very low cost of DTT boxes makes it the ideal means of distributing TV to all the secondary rooms in the house. TV broadcasting is a stretch for a broadband access connection. This is less to do with the speed of connection than the sheer volume of data that has to be shipped to support a typical family TV consumption. It is not just bursts of data but live TV demands a low network 'contention ratio'. Data capacity shipped on this scale translates into huge network investments. Scheduled TV is a fault line in the IP convergence story.
- The emergence of very low-cost bulk storage (cheap high-capacity hard drives) will also have in impact in the light Internet user home. This offers the cellular radio operators the opportunity to use their smaller capacity data pipes to pump out content overnight when their networks are empty.

Almost certainly the competitive challenge of low-cost telephone calls using VoIP over WiFi over broadband will drag down the cost of making telephone calls using the mobile phone to almost zero in the home. Or more likely is ever-bigger bundles of mobile voice minutes tending towards flat rate packages. At this point an inversion takes place where heavy mobile telephone users who were the best customers under a usage model become the worst customers in a flat-rate model (they drive the most cost) and the light usage customers become the best customers (they pay the same flat rate but hardly use the service).

**Homes with no DSL Connection**

There are areas of Europe with relatively low PC penetrations and this effectively puts a lower ceiling on broadband penetration in those markets. As the price of mobile phone calls falls to head off the threat of VoIP over DSL, the non-Internet homes will be incidental beneficiaries. This in turn will knock away the main support for retaining a wire-line telephone. These homes should become the cellular radio operator's mainstay as they reap the 'regulated' access fee those homes were paying the incumbent wire-line operators.

Free-to-air digital TV is the natural complement to the mobile radio services provided to these homes.

## 13.7.4 Convergence and the Evolution of Mobile Handsets

The 'battle for the home' will manifest itself in the future design of mobile phone handsets. Figure 13.10 shows where we have come from and where we have arrived. It has become a classic consumer electronics space driven by an intoxicating (or lethal) mix of fast-moving changes in technology, public taste and fashion.

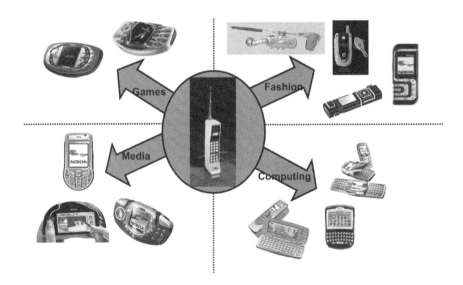

**Figure 13.10**   The evolution of mobile phones.

In the cellular radio industry many cellular radio operators took and retained a sizable hold over mobile phone designs through their control of the retail distribution and handset cross-subsidy. The cross-subsidy model worked well in growing the size of the market and assisting the transitions between mobile phone generations. Today handset subsidies are a millstone around the cellular radio operator's necks and as competition further squeezes margins it is inevitable that cellular radio operators will look for ways to get off the handset subsidy treadmill.

One can see several forces at work that will shape the future of mobile phones.

- The first is the counter-revolution to the 'Swiss Army Knife' approach to mobile phones. The more functions that are packed on to a mobile phone the less likely the phone will be optimal in handling any specific one of the functions. This is likely to be one dimension of the segmenting of future mobile phones.
- The second force is commoditisation. When phones are very cheap consumers will have a number of them. One is likely to be built into the car (no more tedium of changing car kits every time one buys a new mobile phone). Others will be bits of mobile phone functionality built into a variety of other devices.
- The third force is design and fashion as a means for suppliers to add value to a product that otherwise offers otherwise tiny margins.
- The fourth is convenience. Mobile phones have become a vortex sucking in consumer applications such as cameras and music players. Door keys and wallets are good candidates to be sucked into becoming standard mobile phone features using RFID technology.

## 13.7.5 Summary Impact of Convergence as a Disruptive Force

On our 5- to 10-year horizon the most notable thing to any observer in the period 2011–2016 will be an appearance of incredibly fast change – seen primarily through ever-divergent choices of mobile handsets. Beneath the surface will be a demographic and geographic landscape of huge contrasts (see Table 13.3).

**Table 13.3**   Re-shaping the vision 5–10 years out by the forces of convergence.

- WiFi, GSM and digital storage will creep into every nook and cranny driven by price disruption from massive scale economies.
- 3G technology will hold its own for wide area services driven on the industrial momentum it has established.
- Mobile TV will run over different network technologies including 3G. We may see GSM/DVB-H/WiFi combination of handsets challenging 3G for mobile multi-media services.
- VoIP will be taking most of the telephone traffic from the home's connected to the broadband Internet.
- WiFi-based networks will dominate short distance connection in the office, home and dense urban centres.
- Fixed-wire line will disappear in homes not connected to the Internet; i.e. mobile only homes will have increased.
- Small islands will emerge of very high-speed access connections in dense urban conurbations (e.g. flats) and this could be the starting place for the next generation of WiFi running ten times faster than today's WiFi.
- WiFi is also likely to be challenged by a new beyond 3G or beyond WiFi technology. Just as Ethernet went from 10 Base T to 100 Base T and then to Gigabit Ethernet, so wireless will follow but limited to distance of a few hundred metres.

We have painted a three by three matrix with dense urban, suburban and rural defining one axis (based on network economics) and heavy-broadband data homes, light-broadband data homes and no-Internet homes defining the other axis (based on customer behaviour/demand).

The intersection of the heavy broadband data user with the dense urban area will be a blast furnace of innovation and technology advance equating to around 30% of the market. The 'no-Internet homes' will be an oasis of technology stability for cellular radio equating to around 20–40% of the market (varying between European Countries)– with the piece in the middle shading between these two polar positions.

The main impact of convergence will be the Internet taking a high percentage of voice calls in broadband homes and the fixed wire telephone getting pushed out of the 'no-Internet' homes (and a proportion of the broadband-light Internet homes). Some homes will not be connected to the broadband Internet for economic reasons but many will be a matter of lifestyle choice.

## 13.8  The Blindside Forces of Disruption

What we have discussed so far is a possible future scenario landscape shaped by already demonstrated customer behaviour, technology possibilities and rational economic forces. We now look at a list of disruptive forces that can blindside our projected scenarios. Those forces are:

- governments
- regulators
- competitors
- suppliers
- financial markets
- customers
- technological advances.

### 13.8.1  Governments

The whole relationship between the incumbent operators, independent regulators and governments is a delicate one. Regulators are independent because governments want it that way. It takes them out of the firing line of difficult decisions and it went with the flow of economic liberalisation theory of the 1980s. But economic theories can go out of fashion. Over-burdensome labour laws coupled with high levels of debt and a downturn of telephone revenues (impact of VoIP?) can make political solutions look more attractive to incumbents in some European countries than commercial ones. A trigger could be a government wanting investments that market forces alone would not drive and the conditions are there for the 'in' to disappear in the phrase 'independent regulator'.

If the political forces re-enter to 'pick technology winners' it is not hard to predict where they are most likely to exert their influence. Politicians love large infrastructure projects (fibre getting ever closer to the home), they are seduced by television (good news for mobile TV or even mobile satellite TV) and they have elections to win in rural areas (leading to pressures to roll out new networks well beyond dense urban areas).

The front-runner on today's evidence is VDSL to the home. This upgrades DSL access speeds of up to 6 Mbits/s to an eye-watering 25–50 Mbits/s.

The case of Telekom's wish for a regulatory waver in exchange for investing $3.6bn rolling out a more advanced high-speed Internet access services (fibre to the cabinet and VDSL to the home) to 50 Germany cities is an interesting example of the sort of pressures than can shape future infrastructure development.

The impact of this sort of development is not just making 3G HSDPA customer data speeds look tame but WiFi will start to look like a bottleneck in the home. Already WiFi is on a fast evolution path to higher speeds.

More fundamental is that fibre arriving within 300 m of large number of homes starts to create a local backhaul infrastructure that a fourth-generation mobile technology could be dropped on to in dense urban areas.

It is not just central governments with the potential to interference in the market. We already see some city authorities getting into the business of subsidising 'free' WiFi Internet access areas. Rural coverage using more advanced networks is another candidate for governments disrupting the outcome of rational market forces.


## 13.8.2 Regulatory Loose Cannons

The emergence of the independent regulators in the communications sector has been helpful in the globalisation of markets and necessary in limiting the abuses of the former telecommunications monopolies. The theory of independent regulators is that they should provide a predictable environment that in turn encourages investment. However, somewhere in their creation has become a holy writ that independent regulators in Europe should be 'technologically neutral'. Yet many of their decisions impact the environment for a particular technology and the shape of future communication infrastructures. Their studied blindness to the technology impact of their decisions leads to regulators acting like loose cannons for those seeking to advance particular technologies.

One way that technologies may get inadvertently advanced or impeded could result from the quite unequal regulatory oversight of the various converging industries:

- The incumbent fixed network operators continue to be held in a tight grip by Europe's communications regulators by virtue of their dominant position in the last-mile copper wires.
- To a growing extent, Europe's cellular radio operators are coming under the vice-like grip of the Commission and some national independent regulators in spite of the fact that they operate in a fully competitive market. An interesting aspect of the EU Commission attack on international roaming charges of the cellular radio operators is that these roaming charges were one of the important bits of glue that stuck the original GSM initiative together – Europe's greatest success story in high technology.
- The Internet players operate in a largely 'regulator free zone', subject only the general competition law. Yet a few of the Internet giants have almost unassailable dominant positions, for example Google for search-related advertising and eBay for on-line auctions.
- Broadcasting regulation is a whole world of its own with a very strong national character.

This mishmash of regulatory environments will tend to favour the least regulated (the Internet players) and the most regulated (the broadcasters, where regulation gets translated into protection).

This in turn may well create a very global character of the market where the Internet players exert their force and very strong national character where broadcasting players are the force to be reckoned with.

The heart of the issue will come down to bundling and the degree of cross-subsidy allowed across the triple or quad plays. Put another way, which services will be discounted or given away free to attract consumers to a bundled offering of fixed voice, broadband Internet access, TV and mobile? In this game the regulated separation of the incumbent fixed operators' mobile businesses will be a swing issue in the market and could skew the game towards converged offerings from the incumbents or otherwise.

## 13.8.3 Disruptive Competitors

Suddenly the competitive cellular radio market has a lot of new faces appearing from other industries and particularly the large Internet players. Where they could be particularly disruptive is in bringing in new business models. It is hard for a mobile operator to compete by offering the lowest possible price (related to cost) with Internet player giving voice calls away free but effectively cross-subsidised by selling, for example, search-related advertising or the latest new Windows shrink-wrapped software. The most likely impact of the Internet players will be in the messaging space where Instant Messaging communities are likely to span across the fixed and mobile networks as a common space.

'3' is a good example of a disruptive mobile network competitor. What they have done in the 3G market is a classical new-entrant tactic of buying their way into the market by hugely discounting prices. The Italian market was particularly vulnerable to a new entrant introducing a handset-subsidy business model. But in the context of shaping the future of technology it is 3's DVB-H initiative that deserves particular mention. They bought a local Italian TV station just to gain access to the frequency spectrum to roll out their own DVB-H network. That was not an economically rational thing to do. This in turn has pushed TIM and Vodafone into a relationship with Mediaset to lease capacity on their DVB-H multiplexer. This will inevitably lead to a skew particularly in the Italian market for mobile phones with larger screens.

A third area of irrational competitive behaviour would be to see a 'comparative competition' between incumbent fixed operators rolling out VDSL Internet access. The existence of a massive over-supply of fibre optic cable in a number of the larger European cities offers the fixed infrastructure to roll out WiMax services. Fibre to the apartment is well within reach of affordable technology. A gigabit Ethernet router in the basement of a high-rise flat connected to fibre in the street, multimode fibre run up the lift shaft to each flat and a media converter gives customers 10–100 Mbits/s access rates to the Internet. In fact all the large cities in Europe that are fibre-rich are ripe for an irrational technology push 'arms race' towards an order of magnitude increase of Internet access speeds.

## 13.8.4 Disruptive Suppliers

Intel is a good example of a 'disruptive supplier' with their WiMax initiative. Had a giant of Intel's size not pushed the WiMax technology, it certainly would not have got as far as it has on its own merits. The sheer marketing 'noise' (which I distinguish from information) that Intel has managed to generate and the support it got from US trade diplomats pushed European governments to make spectrum available.

'WiMax has a useful future as a mainstream global wireless local loop technology but is future as a global mobile technology is far more problematic. It missed the time window for a global $3^{rd}$ generation cellular technology and is too soon for a $4^{th}$ generation cellular technology in the natural order of things. It will cost a lot more to roll out any sort of wide area mobile coverage at 3.4 GHz and HSDPA is rapidly closing down the space that might have existed in many markets. Huge investments will be needed to pull it out of its unvirtuous circle and even then it is not obvious why it would achieve competitive industrial parity with 3G UMTS.'

## 13.8.5 Gyrating Financial Markets

Whilst memories may be short, we hope that they are not short enough to see any early return to the destructive exuberance for the telecommunications industry of the financial markets in 1999–2000.

That said, there are huge amounts of capital searching for opportunities to bring superior returns and no shortage of entrepreneurs with dreams.

There has also been a lot of quiet activity around the fibre assets of entities that went bankrupt in the telecommunications crash of 2000. Enough restructuring of companies holding these long-distance transmission assets might firm up prices that, in turn, could put constraints on the rising amount of video traffic across the Internet or even shake the grip of the 'flat-rate' pricing model for broadband access.

The real signal of financial market exuberance returning would be a revival in mobile satellite projects.

## 13.8.6 Unpredictable Customers

SMS is often quoted as the best example of unpredictable customer behaviour. What started out as an offering intended for the business market as a better form of paging messages suddenly erupted as a school playground phenomena. From there, SMS has progressed on to being a very important secondary means of mobile communications and a huge revenue earner for the mobile operators. The ring tone market has been another pleasant customer behaviour 'surprise' for the mobile network operators. Areas to watch are:

- What may suddenly sprout from the vast amounts of private data being captured by millions of cameras on mobile phones and PCs?
- Will HDTV take off in Europe? HDTV is massively data hungry and if it got any serious traction in the consumer market it would make IPTV across the broadband Internet less viable/attractive. Arguably much of the picture quality enhancement could be done through processing in the TV receiver without the need to transmit all the extra data. Most consumers would not notice the difference between a progressively scanned TV picture that is transmitted at standard definition and line up converted in the TV set and a transmitted HDTV picture. But marketing departments are adept in helping consumers think otherwise.

- Customers may wake up to what data speeds most of them actually need and not be mesmerised by the headline data speed caps that in practice they rarely ever experience.

## 13.8.7 Disruptive Technologies

Fibre optic cable remains potentially transformational for the whole telecommunications industry. The extent to which fibre cables are brought to within 100–300 metres of peoples' homes will determine the viability of a massive upgrade of wider area mobile radio data speeds.

Fibre to the home is a dream that has been around for a long time. In the 1980s, HDTV was thought to be the killer application to drive it into the market. At that time HDTV demanded a phenomenal 140 Mbits/s. Then GI showed the world how it could be done in 15 Mbits/s and that market driver for fibre to the home was derailed. Meanwhile technology advances with data over copper has been driving forward apace. Whilst the conditions are right for fibre to the apartment (needing only a few really data-hungry applications to become attractive) they do not look right for a mass roll-out of fibre to the home, at least over the next five years.

Data storage technology is another massive disruptor. It is not that difficult to arrange for the mobile phone battery charger to also become a 'data file' charger and a massive amount of data traffic that could have passed over mobile networks will have been redirected over the fixed broadband Internet. Most sound radio stations are little more than music 'play lists' interspersed with news flashes. The music files could all be preloaded in the mobile phone and all that needs to be transmitted is the play list, news flashes and the occasional new release.

Waiting in the wings for the cellular radio industry are some technologies that perform better than 3G UMTS. A notable one is OFDM (orthogonal frequency division multiplex) technology that is already in widespread use for digital terrestrial television. It was a candidate technology for 3G but there were problems for its application for mobile that have since been solved. It therefore missed the bus. It is not easy to see a plausible scenario that will create the conditions for its mass deployment in Europe over the next five years.

Finally, intelligent antennas at either end of the mobile radio link could transform data speed, capacity and distance constraints for cellular systems but the costs have to be tamed to see wide commercial deployment.

## 13.8.8 The Global Perspective

The analysis so far has been European centric. In the 1990s, GSM catapulted Europe into the world leadership position in mobile radio. UMTS has not been a re-run of the same success story. There were a number of strategy errors made:

- There was a miscalculation by the large European suppliers who drove the 3G industrial strategy towards a link up with Japan rather than building out from their GSM strength.
- A 5 MHz wide channel did not allow the technology to be weaved into the GSM spectrum very easily.
- The suppliers failed (until very recently) to deliver its original promise of 2 Mbits/s customer data rates. And what was delivered was late. On both counts this allowed in WiFi to fill some of the customer needs (nomadic laptops).

- There was carelessness on everybody's part in failing to tie down the precise terms of licensing intellectual property rights at the time when the standards body (ETSI) had the choice of walking away and selecting an alternative. The cumulative IPR royalties on 3G technology has been a substantial millstone around the 3G technology.
- The astronomical cost of 3G licences was another misfortune. Imagine the impact if even half of this licence fee had been put into the 3G infrastructure roll-out itself?

Whilst 3G is now well passed its tipping point, Europe blew away 'an unassailable' global leadership in mobile radio. It is now a much more level playing field of leadership opportunity. The USA has undoubted leadership in the Internet and IT spheres and we can expect this industry and local market to be out front for this technology track. Very often Asia is characterised as one homogeneous region. In fact the dynamics in Japan, South Korea and China are very different. There is much more evidence of a 'Korea Ltd' than we see in Japan. Japan has a hugely fierce domestic market that seems to have an insatiable appetite for new technology wizardry. China has an awesome combination of a huge domestic market and growing technology capability – but could misfire on the subtle strategy needed to harness competition in pushing out new technology into a global market. The key is what is unique to that market that gets commoditised and tumbles into Europe driven by its low cost. Even then a global oriented strategy is essential from the outset if the lack of success that occurred with some second generation digital mobile standards in the Japanese and US markets is to be avoided.

The emerging markets are attracting a lot of global investment in cellular radio networks. The growth potential here is enormous. A number of these countries need a telecommunications infrastructure as a prerequisite to developing effective local markets for the wider economy of goods and services. Almost certainly mobile radio is the most cost-efficient way to go. Faced with the choice of a $20 mobile phone or a $100 PC the choice is self-evident in terms of economic development.

In innovation terms we should not write off emerging markets as just recipients of hand-me-down technologies from the developed world. Already we have seen innovative banking services appear based on the mobile phone pre-paid SIM card filling a banking system vacuum. Even on the hardware side it may not be surprising to see $40 mobile phones that expose enough PC processing power that can be linked to cheap hard drives, keyboards and displays in due course.

The global picture is of a diverging mobile technology world over the next five years that will act to strengthen the status quo in Europe, at least at the network infrastructure level.

### 13.8.9 Summary Vision of the 'Blindside' Forces

Table 13.4 re-shapes the vision 5–10 years out by 'blindside' forces.

## 13.9 Conclusions

We started our projection based entirely on what appeared to happen in the past. This would have seen a fourth-generation cellular radio technology being introduced around 2014 that would displace GSM and eventually displace 3G. It would be an exercise of 'game theory' between governments, network operators and the supply industry. We then looked at the

**Table 13.4**   Re-shaping the vision 5–10 years out by 'blindside' forces.

- Dense urban areas could become a bloodbath for the network operators as prices fall under intense competition and costs rise due to irrational technology investments including VDSL, WiMax and the municipalities rush in to offer carpets of free WiFi access.
- The single European market in telecommunications fractures into a series of very distinct national markets shaped by the results of nationally centric broadcasting, legacy centric fixed telecommunications operators, global Internet players and the influence of governments/regulators.
- National markets fracture in turn according to customer behaviour towards the broadband Internet and in particular whether it is central to their lives, an adjunct to it, or plays no part.
- Satellite mobile services driven by irrational investors supporting disruptive suppliers.

future from a theory that large networks have momentum and just carry on evolving unless disruptive forces act to change that state. This blew away our assumption that GSM would be closed down by 2014. Then we examined various disruptive forces we knew about and convergence in particular. This clouded the picture. Other technologies come into view that will contest the role of 3G technology. Mobile radios as we know it may get pushed entirely out of some homes but entirely take over others.

All these various forces we have looked at attack the assumption that infrastructure investment in the European cellular radio market will occur in concerted well-behaved 12-year cycles. The probability is high of many different technologies being tried in the market and quite low that any of them will achieve a virtuous circle. If a fourth-generation technology emerges with the same momentum as 2G and 3G it will be more a matter of happenstance than something that is planned for. Our cellular radio world has become 40% 'game theory' and 60% 'chaos theory'.

## Biography

*Stephen Temple CBE, MSc, CEng, FIEE*
Stephen Temple joined Vodafone's Corporate Strategy Group as Director of Strategic Projects in July 2003. He spent seven years with the UK's largest cable TV operator NTL, for the most part leading their technology innovation programme in the Networks Division and finished up as Managing Director of the Networks Division. Prior to that he was a senior official in the Department of Trade and Industry leading for the UK in a number of European telecommunications developments

including the GSM mobile project, digital video broadcasting, collaborative R&D and telecommunications standardisation. On the latter he was Chairman of the ETSI Technical

Assembly for four years. In an earlier part of his civil service career he spent seven years in frequency spectrum management in the Ministry of Posts and Telecommunications and Home Office. He was awarded the IEEE prize for International Communications in 1994, the GSMA chairman's award in 1996 and the CBE for services to Trade and Industry in 1996. He is a Fellow of the Institution of Engineering and Technology (IET; formerly the IEE).

# 14

# Assimilating the Key Factors

## 14.1 Introduction

This chapter is where we bring together all the key concepts, thoughts and conclusions that have been mentioned in the book so far, both in the first part and by the contributors. This will then provide us with a set of 'interim conclusions' as a basis for predictions in the next chapter.

## 14.2 Summary of the Current Position

When considering the current environment we conjectured that:

- 3G will be increasingly deployed, becoming the key cellular system within the next five years. 2G will continue to exist in parallel for at least another decade. Cellular operators will enter a long period of stability.
- WiMax is unlikely to make a major impact on the success of 3G.
- There may not be a discrete 4G, based on a new standardised technology; instead there will be incremental improvements in existing standards and their interoperability.
- Convergence between cellular and wired networks will likely be based on W-LAN in the home and office, and cellular in the wide area.
- Mobile mesh systems do not have sufficient advantages to bring about their widescale deployment but may have a plethora of niche applications.
- Fixed wireless will likely not succeed substantially, even with the introduction of a nomadic offering.
- The key short range devices will be W-LAN providing in-building networks and BlueTooth providing device-to-device connectivity.
- Core networks will migrate to IP over the next decade.
- Conventional broadcasting will continue long into the future.

*Wireless Communications: The Future*   William Webb
© 2007 John Wiley & Sons, Ltd

- Building separate networks for mobile broadcasting seems hard to justify although it is difficult to predict demand.
- The current industry structure is for providers to be vertically integrated. This will slow the speed of convergence due to the need to form partnerships rather than just procuring services.

In short, our view of the current situation was that it was a stable one. The major operators were well established with clear plans. New technologies were unlikely to make much impact or destabilise the situation, although some convergence might occur. But our expectation for significant change was low.

## 14.3 Summary of End User Demand

When assessing end user demand we concluded that:

- For a new service to be accepted it must be something that the users actually want. However, users often cannot envisage future possibilities and needs, so simply asking them might not deliver the right answer.
- Users take time to adopt new services and ideas. It can take between four and ten years for even the most 'perfect' service to become widely adopted, and there are many pitfalls on route that can extend this, or even cause the service to fail.
- Users have limited resources to spend on communications and take time to change their spending habits. Any expectation of revenue growth in excess of around $2.80/month happening in less than a year is likely to be unrealistic, unless there are unusual circumstances or conditions.
- Users today have a wide range of communications offerings, but this range itself causes problems of multiple numbers, mailboxes, etc. Simplification of this would bring benefits.
- Services that would bring benefits to users in general are those that make communications ubiquitous, simple, inexpensive and unrestricted in format, data rate etc.

In summary, the key message was that even the most perfect service takes some time to diffuse across the population, limiting the speed of change.

## 14.4 Summary from Technology Advances Section

Summarising the section so far:

- Technical efficiency of wireless systems is approaching fundamental limits. While there is some prospect that MIMO antennas might increase capacity beyond these, this prospect seems relatively slim, especially as their benefits decline in a small-cell environment.
- The many empirical laws suggest that capacity and data rates on wireless networks will continue to increase but primarily driven by ever smaller cells.
- There are few technologies currently under investigation that promise any dramatic change in the technical efficiency of wireless systems.
- Display technology on mobile phones will steadily improve and network-based speech recognition will bring significant gains, but interacting with the mobile will remain compromised compared to devices such as PCs.

- Steady progress is expected in areas such as batteries, processing power etc.
- Backhaul becomes one of the key constraining factors and the biggest challenge facing wireless communications.

All these conclusions suggest that there is no wonderful technology, or technical trend, that on its own is going to revolutionise wireless over the next decade, and with all likelihood over the next 20 years. Equally, we can expect increased capacity and data rates almost entirely as a result of smaller cells. These might be a mix of cellular networks continually reducing cell sizes in urban areas, and the increasing deployment of W-LANs, in some cases providing coverage across entire cities.

## 14.5 Summary from the Contributors

The following paragraphs provide my summary of what I feel to be the key factors brought out by the contributors, and not their own summary (which they generally provided at the end of their chapters). To be clear, the material from this point on has not been seen by nor agreed with the contributors, and is the work of the author alone, based on all the conclusions drawn so far.

### Paul Cannon

- There are some fundamental differences between military and civil communications such as the need to deploy quickly and not to rely on single points of failure. However, there are also many similarities.
- Military operations are changing and hence the wireless requirements are changing too.
- Being able to interwork with civilian systems during operations such as disaster relief is helpful. This might require SDR functionality.
- HAPs are likely to be deployed for military usage, and as a result may subsequently make it into the civil world.
- Body area networks are strongly needed for soldiers. Again these could be useful to the civil world.
- The military is driving research into low-cost, small sensors – another area of clear civilian need.
- HF communications needs are driving lower voice compression rates, but the quality may be insufficient for civilian use.
- Civil applications will only use mesh as 'stubs' to existing networks, but the military will require larger and more dynamic mesh networks.
- SDR and cognitive radio remain important concepts for military operations.
- Propagation research, especially in buildings, may lead to better models over time.
- Advanced satellite systems will eventually be deployed at higher frequencies, but it is unclear as to whether there is a link to civil applications.
- Nanotechnology and novel power sources may bring some new properties, particularly to sensors.
- Overall – a prediction of a huge increase in data throughput for future military systems compared to those of today.

## Peter Cochrane

- Management of the radio spectrum is no longer needed. It can be opened up to allow people to do what they like.
- Spread spectrum and UWB will provide 1 Gbits/s data rates but over <30 m.
- Plug and play will become increasingly effective and expected.
- More wireless means more fibre to connect up the cells – which in turn drives wireless and wired companies to combine. With fibre, direct radio may become practical.
- The home will change dramatically as hundreds of wireless transmitters are deployed perhaps using 60 GHz units.
- RFID will dramatically reduce overall transaction costs.
- Wireless will be used in transport to track containers and vehicles leading to high capacity radio nodes being deployed in key locations such as petrol stations.
- Parasitic networks will develop to extend coverage.
- Sensor networks will become widespread allowing monitoring of many environmental and other factors.
- The future is an unusual but beneficial wireless cloud.

## Gary Grube and Hamid Ahmadi

- The cost and difficulty of providing wireless broadband will increase exponentially from today's model.
- Ad-hoc networks minimise backhaul and make it economic to extend coverage into urban and suburban areas.
- Sensor networks will become increasingly important across all aspects of life and ad-hoc is ideal to support this. Homes will keep track of people and temperature. Sprinkler systems will know the condition of the lawn. Roads will have sensors in the lane paint to detect vehicle movement.
- There are many 'special cases' where mesh is perfect – such as emergency services.
- Low-power photocells will allow sensors to work without batteries.
- Ad-hoc networks scale down as well as up.

## Dennis Roberson

- Regulation which has been helpful in the past is now an encumbrance to new emerging applications.
- Unlicensed spectrum will become overwhelmed by a 'perfect storm' of growing device numbers, data rates and applications.
- The best approach to solve spectrum shortages is cognitive radio.
- WiFi deployments will become increasingly ubiquitous, wherever people gather.
- New applications will include large-scale multi-user games and multi-authoring activities.
- Gaming will expand to include three-dimensional capability and touch.
- Better realism in communications will enable, for example, a family to share in the travel experiences of one member.

- In 2025–2030 we will see the emergence of 'mind-to-mind' communications including the ability to control electronic things in our daily lives. Robots will also emerge around this time.
- Managing the spectrum utilisation to dramatically enhance its data carrying capability is the most critical piece of this future.

## Simon Saunders

### Predictions for the period 2006 to 2011
- There will be rapid growth in demand and devices.
- In-building systems will be slow to emerge due to interference and other problems and cellular operators offering low tariffs to fight back. There will be widespread use of both WLAN and 3G 'femtocells' in offices.
- Service providers will deliver home routers to help set up home network systems.
- Mobile TV will be widely used but over 3G.
- City-wide WiFi networks will be unsuccessful.
- Consumer brands start to own wireless networks in their buildings and work with others to deliver consolidated services.

### Predictions for the period 2012 to 2016
- Devices span all relevant technologies.
- Machine-to-machine data exceeds personal data.
- Buildings become increasingly wirelessly controlled.
- MIMO becomes widespread in devices.
- Service provision becomes separate from network ownership, and other models such as aggregators emerge.
- Backhaul becomes a key constraint.
- Meta air-interfaces emerge allowing different standards to co-exist. No new standards are developed.
- Spectrum trading abuse causes a crisis of confidence.

### Predictions for the period 2017 to 2026
- Domestic automation becomes standard, but took longer than many expected. Security issues are an initial problem.
- Voice recognition becomes the input mechanism of choice.
- Network operators and service providers separate.
- High performance is delivered by complex multi-base station MIMO systems.
- Smaller cells provide one to two orders of magnitude gain in capacity.

## Stephen Temple

- Large networks are either in an unvirtuous circle or a virtuous circle. Very large forces are needed to flip from one to the other. It is difficult to get a new network technology off the ground.
- Specialised mobile networks such as WiFi and DVB-H will sit alongside 3G, not beyond it.
- If there is a 4G it will be targeted at dense urban areas only, resulting in tiers of coverage of different generations out to increasingly rural areas.

- WiFi will be used in the enterprise and home for convergence of voice and data.
- GSM will continue to exist and may have a key role in machine-to-machine communications.
- Mobile TV may emerge in many forms depending on demand, screen size, use of podcasts compared to live viewing, and may only be in urban areas. It may run over a range of dedicated and shared networks.
- 3G will be used for wide area distribution.
- VoIP will be widely used in the home.
- Provision to the home is very dependent on customer preferences and behaviour. Mobile-only homes that dispense with a fixed line will increase. Also there will be heavy users that have limited requirement for mobile.
- Very high-speed connections will emerge in small localities such as flats in dense urban areas.
- 4G will eventually challenge WiFi.
- Intense competition could arise in urban areas due to irrational investment and local authority behaviour.
- Lots of technologies will be tried in the market but the probability that any will succeed is low and will be more by accident than design.

## 14.6 Key Factors brought out by the Contributors

The contributors have raised a large number of important and insightful points. Pulling out the most important of these is fraught with difficulty, but necessary in order to bring out the key factors. In this section, we draw out those areas where, compared to the first seven chapters of this book, the contributors have identified different areas of importance, or disagree with earlier findings. We have not drawn out conclusions relating to those areas where they broadly agree with the first seven chapters. These areas of agreement are by far the most numerous, suggesting a good level of confidence in the bulk of the future predictions.

### 14.6.1 Areas not Included in Previous Discussion

**Connectivity**

- Spread spectrum and UWB will provide 1 Gbits/s data rates but over <30 m.
- Wireless will be used in transport applications to track containers and vehicles, leading to hotspots in key locations such as petrol stations.
- Many different types of provision to the home and distribution within the home will exist depending on the demands and preferences of the occupants.
- City-wide WiFi networks will not be successful.
- Consumer brands, such as supermarkets, will start to deploy wireless in their own buildings and form consortia with others to turn this into a useful wireless resource.
- Meta air-interfaces will emerge allowing different standards to co-exist. No new standards are developed.
- Mesh or ad-hoc networks (also termed 'parasitic networks') will emerge as a key facilitator for wireless connection across a wide range of applications.

- In 2025–2030 we will see the emergence of 'mind-to-mind' communications including the ability to control electronic things in our daily lives. Robots will also emerge around this time.

**Backhaul**

- More wireless means more fibre to connect up the cells, which in turn drives wireless and wired companies to combine.

**Applications**

- Sensor networks will become widespread, allowing monitoring of many environmental and other factors.
- New applications will include large-scale multi-user games and multi-authoring activities.
- Gaming will expand to include three-dimensional capability and touch.

**Technology**

- Military systems might provide a driver for a range of new technologies including SDR, HAPs, body area networks, low-cost sensors, mesh and cognitive radio. This may speed the introduction of these technologies into the commercial world.
- Very large forces are needed to get a new network technology off the ground. Many different technologies will be tried but few will succeed.
- Machine-to-machine communications will exceed person-to-person communications by around 2012.

**Regulation**

- Spectrum trading abuse causes crisis of confidence (but ultimately leads to better regulation).
- Regulation which has been helpful in the past is now an encumbrance to new emerging applications.
- Managing the spectrum utilisation to dramatically enhance its data carrying capability is the most critical piece of this future.

## 14.6.2 Areas of Disagreement

- MIMO will become widespread in devices around 2012 and networks around 2017.
- Unlicensed spectrum will become overwhelmed by a 'perfect storm' of growing devices, data rates and applications.
- The best approach to solve spectrum shortages is cognitive radio.
- WiFi deployments will become increasingly ubiquitous, wherever people gather.
- Management of the radio spectrum is no longer needed. It can be opened up to allow people to do what they like.
- Mesh networks will be the key driver for wireless growth in the coming years.
- Direct radio will become possible and advantageous.

## 14.7 Reaching a Verdict on the Areas of Disagreement

The areas of 'disagreement' show where there are areas of significant doubt about the future. Because the aim of this book is to drive towards a single future roadmap, this section discusses these areas and reaches a verdict. It is the verdict of the author and not of the eminent contributors and so must be treated with some caution.

1 *MIMO will become widespread in devices around 2012 and networks around 2017.* A number of contributors predicted this. However, we were of the view that smaller cells would reduce the benefits from MIMO and generally render it uneconomic. Having said that, given that MIMO will lead to much smaller capacity gains than those from deploying smaller cells, predicting MIMO penetration correctly is only of low sensitivity. Hence, we can sit on the fence regarding this one – it will not materially change our predictions.

2 *Unlicensed spectrum will become overwhelmed by a 'perfect storm' of growing devices, data rates and applications.* There is substantial doubt as to whether unlicensed spectrum will be able to cope in the future – there is not even much evidence as to its levels of congestion at present. In considering the way this might develop it seems likely that congestion will be localised – congestion may occur in an airport lounge, but not in the nearby car park, or in a downtown office but not a nearby warehouse. Congestion will also build gradually over time, with achievable data rates declining as more users make use of the spectrum. This suggests that manufacturers and users will be able to change their devices and behaviour over time rather than suddenly facing catastrophic failure. They may move to different frequency bands, adopt different protocols or even move to different building materials that restrict wireless propagation. More spectrum can also be made available. Given the importance of this area it seems likely that solutions will be found to prevent a major failure.

3 *The best approach to solve spectrum shortages is cognitive radio.* We have discussed cognitive radio at some length. While there is certainly significant under-use of spectrum it is not clear whether cognitive devices will work, or are the best way to make better use of spectrum. The technology is clearly some way off and will be more expensive than conventional handsets. Also, history suggests that using smaller cells will continue to provide much greater growth in capacity than cognitive radios could. So on balance, we do not think cognitive radio will be a key driving force within the next 20 years – but others, who have excellent credentials, do. This is clearly a point of importance.

4 *WiFi deployments will become increasingly ubiquitous, wherever people gather.* Between the contributors there is divergence around the extent to which WiFi will be deployed. Some believe city-wide deployments are likely, others that these will fail. Some see mesh technology as a way to extend coverage into suburban areas. The density of high-speed wireless nodes will have a major effect on the data rates and facilities available to end users. But perhaps if we delve deeper there is less difference. All agree that WiFi will be deployed in dense areas and also areas where people congregate – coffee shops, sports stadia etc. Mesh will likely allow the extension of WiFi to other areas, but with lower capacity and data rates because of the problems with mesh backhaul. Hence, we will see a gradation of WiFi capabilities from high data rates in urban areas to lower rates and more sporadic coverage in suburban areas. This is also likely to be true of city networks which will offer very mixed coverage and capacities throughout the city.

5 *Management of the radio spectrum is no longer needed. It can be opened up to allow people to do what they like.* This view is linked with technologies like cognitive radio that would allow management of the radio spectrum to be reduced. As we have argued above, the future for cognitive radio is not clear. Also, to date, the only significant investment in wireless infrastructure has been in licensed spectrum. Investment and revenue from licence-exempt spectrum is typically less than 1% of that in licensed spectrum. Given the time it takes for investment to occur in networks and the doubts about the technology, it seems unlikely that spectrum management can be withdrawn for at least ten years, maybe longer. However, it should be possible to progressively reduce the amount of regulation needed, relying more on the market to find the best use for the spectrum.

6 *Mesh networks will be the key driver for wireless growth in the coming years.* One of the chapters argued eloquently and at length as to why mesh architecture would make a major difference to future of wireless. In other places, we have argued that mesh has many difficulties including less capacity than equivalent conventional networks, increased delay and much increased complexity. However, the chapter on mesh showed how it could be used effectively in a number of situations where capacity was less important than providing connectivity which otherwise might not be available. We would therefore conclude that mesh will not change the 'mainstream' provision of wireless coverage but will be critical in many applications such as emergency service use, the home, sensor networks and potentially networks between cars on the road.

7 *With fibre, direct radio may become practical.* As we discussed in Chapter 6, while direct (or fibre) radio remains a tantalising prospect, in our view, as backhaul becomes the biggest constraint, systems such as direct radio that require substantial backhaul resources will not prove successful.

## 14.8  Drawing these Key Factors Together

Summarising all these trends, we have so far said that:

- Data rates will continue to grow on both fixed and wireless networks. On fixed networks this will be achieved by rolling fibre deeper into the network. In mobile networks it will be achieved through the deployment of ever smaller cells. A key limiting factor on the deployment of smaller cells will be the availability of low-cost ubiquitous backhaul.
- Advances in handsets will be held back by the slow developments in display technology and battery power, although speech recognition may advance relatively quickly. Handsets will gradually gain ever increasing functionality.
- User acceptance of new concepts will be slow, and will depend on the cost being reasonable. This will make, for example, remote-control central heating systems very slow to emerge.
- Home wireless networks will become widespread and increasingly important for the functioning of the electronic systems in the home.
- Broadcasting will slowly evolve towards a personalised model, but this will take decades.
- Making the mobile phone context-aware and intelligent is proving very difficult. More likely the intelligence will reside in the network with simple commands sent to the phone. Progress will be slow and limited.

- Key areas of doubt include the extent of deployment of cognitive radio, MIMO and mesh networks, the need for future management of radio spectrum and the extent of deployment of WiFi networks.

A key conclusion is that change will be slow. This is due to the current situation being stable and profitable, the slow pace of technological change, the long time taken for users to change behaviour, the current vertical structure of the industry tending to act against convergence, and the difficultly in making key advances. Indeed, a slow pace of change has been occurring for some years. As the evaluation of the prediction made in 2000 for 2005 showed, the assumption that very little will change delivered accurate results.

# 15

# The Future Roadmap

## 15.1 Introduction

In this chapter we pull together all the material from earlier chapters to make our predictions for the next 20 years from the date of writing (2006), predicting at five-year intervals. We have chosen not to provide a range of scenarios – this just puts the onus on the readers to decide on which scenario they consider most likely. However, where events could change our predictions considerably we indicate what these might be and the changes they might lead to. After setting out our predictions, some of them illustrated with 'a day in the life' examples, we discuss the implications for stakeholders throughout the industry and compare this forecast with our previous one, analysing the differences.

These predictions are the work of the author and have not been seen or agreed with the contributors who provided Chapters 7–13. Having said that, the predictions below were, of course, influenced by the work of the contributors.

## 15.2 Predictions for 2011

### Fixed networks

Fixed networks, with their massive buried infrastructure, change only slowly. The key changes that we predict include a transition to IP in the core, an upgrade of ADSL rates as ADSL2+ is deployed and eventually increasingly deep penetration of fibre into the network. There is an analogous transition for cable networks, towards IP cores, higher speed modems and deeper fibre penetration. None of this will happen quickly. By 2011, BT in the UK may have completed their 'twenty-first century network' project and have implemented one of the first core IP networks. Other fixed operators will be following, in various stages of transition. ADSL2+ will be widely deployed, although the benefits will accrue only to those relatively close to the local exchange. Fibre penetration will be increasing but only substantially in a few select countries where there are strong drivers for fibre. These include Korea and

*Wireless Communications: The Future*   William Webb
© 2007 John Wiley & Sons, Ltd

Japan where there are government incentives, and the USA where competition is spurring deployment. On average, around the world, the end user experience will be somewhat higher data rates, perhaps heading towards 10 Mbits/s for those in urban areas, with the current trend for slowly reducing prices continuing.

### Broadcast networks

As we have noted earlier, broadcast changes only slowly. In 2011, most will still be watching 'conventional' TV broadcasts either wirelessly, via cable or satellite. Viewing habits will be changing somewhat as PVRs become widespread and viewers become more inclined to 'see what the PVR has recorded for me' rather than asking 'what's on TV tonight?' High-definition broadcasting will increasingly become standard for films and sports events. We predict that Internet TV will not have gained significant traction by 2011. Although the required technology such as sufficiently high-speed broadband connections and media-enabled PCs will be widespread, the cultural change needed and the long lifetime of conventional broadcasting receivers will result in most people staying with conventional systems. Mobile TV will still be emerging in 2011. Standards battles, a lack of radio spectrum and a lack of clarity as to the offering will mean that few handsets will offer TV reception as standard and the business cases will still be unclear. By 2011, UHF spectrum will become available across much of the world as analogue TV is switched over to digital and early mobile TV networks will be emerging. The end user experience of broadcasting will be an improved one but due almost entirely to the ability of the PVR to select programming for them.

### Wide area wireless

By 2011, 3G will be by far the most widely used mobile network. 2G networks will still be used – particularly for machine-to-machine communications – but 2G-only handsets will be available only in developing countries. With HSDPA deployed, 3G networks will be routinely offering data rates in the low hundreds of kilobits/s even when cells are heavily loaded, making them acceptable for a variety of data applications. As discussed earlier, WiMax will not have made any significant impact on 3G and no other technologies will be emerging to challenge it. Users will gradually embrace services other than voice, with mobile Internet browsing increasing, especially as more sites become mobile-friendly. Video calling will also be used occasionally as the presence of capable handsets becomes ubiquitous. However, services such as music download will have less success as users realise music is better downloaded at home over a broadband connection and then transferred to the handset. From the end users' point of view little will have changed other than improvements in Internet access and the ability to make video calls should they choose.

### Short-range wireless

Short-range wireless will continue to be dominated by W-LAN and BlueTooth, both of which will evolve to higher data rates, although for most, the data rates available in 2006 were entirely adequate.

Outdoors, hotspot growth will continue between 2006 and 2011 at a rate of perhaps 10–20% per year in terms of number of hotspots. These will expand into more hotels, sports stadiums and to a limited degree covering some cities, although such deployments will not work well at this stage. A number of cellular and fixed operators will start to offer hotspot access as part of a bundled package and cellphones/PDAs able to work on WiFi will become increasingly widespread.

Indoors, this will be a period of significant growth for W-LAN. Home routers with W-LAN capability will become widespread. In some cases these will be provided as part of a convergence package, in others as standard part of a home PC, or media centre. By 2011, over 25% of homes in the developed world will have home W-LAN capabilities, although not all will be actively used. Similar growth will occur in enterprises, with perhaps 25–50% of all businesses having W-LAN access across major parts of their premises. During this period some will start to use the W-LAN for voice (see convergence, below). Other devices will start to make use of this wireless capability. Initially this will be computing devices such as printers, then entertainment devices such as remote speakers, 'Internet radios', etc. Towards 2011 we will start to see home security systems having a W-LAN link to allow remote alerts and interrogations.

BlueTooth will continue to be widely deployed in cellphones and in cars as a means to provide handset-free connectivity. It will also becoming increasingly standard in PCs, starting a trend for devices such as iPods and digital cameras to have BlueTooth capability to allow cableless links to the PC.

UWB might emerge during this period as a mechanism to link video devices such as screens, PVRs, DVD players and speakers together. However, we expect this to be a relatively niche application linked to high-end equipment.

The impact for the end user could be significant. With indoor wireless coverage becoming ubiquitous, a wide range of new applications and convergence opportunities open up.

### Handsets

Handsets have been gradually evolving over time and this process is set to continue as users show an appetite for periodic replacement. One of the key trends will be the increasing addition of functionality. Cameras in phones will progressively improve. Music player functionality will become standard with the memory in phones increasing massively, either via lower costs solid-state memory or through better and smaller mini hard disks. Video storage and replay will also be added towards the end of this period. Speech recognition functionality will slowly be added, in some cases network-based, and more advanced keyboards and capacitive input surfaces will gradually improve interaction with devices. Positioning capabilities will improve, but be restricted to a growing subset of devices due to the cost and added bulk of a GPS receiver unit. W-LAN and BlueTooth will become increasingly standard during this period in handsets.

### Services

Few completely new services will emerge in this period. However, as discussed above, video calling will become increasingly attractive as the 'Metcalfe rule' applies and the network of video-enabled devices grows. Mobile TV will become available towards the end of the period.

### Convergence

This period will be a significant one for convergence. With home and office W-LAN networks becoming standard, devices increasingly able to work on W-LAN networks becoming widespread and core-IP networks coming on stream, all the technology and infrastructure will be available to offer converged voice and data services across the home, office and wide area. The extent to which these services actually emerge will be set by industry structure and the 'boldness' of the strategies deployed by key players (see industry structure, below). By 2011, around 20% of users will have a single handset that they use in the wide area and home, and to a lesser extent in the office, providing a converged voice service. Convergence between telecommunications and broadcasting will be much less advanced, with some viewing of 'podcasts' on mobile devices and limited mobile TV or download of real-time broadcasting but otherwise broadcasting being seen as a separate application.

### Industry structure
Industry structures take a long time to change, particularly where companies are large. However, once a 'tipping point' is finally reached, change can ripple through the industry relatively quickly. This makes forecasting industry structure difficult. Our view is that by 2011, there will be few major changes. Cellular, fixed and cable operators will still be running networks and providing customer service in a vertically integrated form. There will be many deals struck so that, for example, cellular customers can 'roam' on to their in-home network and make calls via the fixed network. By 2011, a few 'integrated' companies will start to emerge through merger and acquisition. For example, cable operators might buy into cellular operators, or vice versa. Towards the end of this period we will start to see the increasing presence of 'service provider only' organisations. These might be entities like AOL, Google, Walmart or others who offer a complete communications package. Emergence will be slow because the vertical operators will fight their presence by not allowing easy access to their networks. However, if these service providers gain traction, it will rapidly move the industry towards a 'tipping point'.

### Other significant issues
Health concerns will continue to periodically arise. However, most research will indicate that health concerns are minimal or non-existent and the increasing evidence that long-term users are unaffected will reassure most. Mast siting will continue to be difficult, but with the trend towards smaller cells, deployments on buildings will be more commonplace, causing less concern.

### End user experience – how will it have changed?
The key change for the end user in this period will be the convergence between home and wide area phones. Users will gradually stop using their home phone, reserving it for those times that their cellphone batteries are low. This will lead to an understanding of the problems of convergence, such as the difficultly in separating work and business calls, and will result in the emergence of filtering solutions to deal with this. Wireless homes will offer a range of capabilities to control the home which users will increasingly come to rely on. This will be the period that many will make their first video call and will start to appreciate that there are times when video calling is advantageous. Users will appreciate higher data rates to the home and in the wide area but will not yet have any new applications that take advantages of these. They will also become accustomed to the mobile phone being a multi-purpose device, with camera functionality acceptable for most and music capabilities sufficient to displace iPods and similar devices for many.

## 15.3  Predictions for 2016

### Fixed networks

By 2016, advances in data rates over copper will have ceased. Data rates will have peaked at around 10–20 Mbits/s for most. Many will see this as adequate and the pressure for higher rates at this stage will be minimal. However, the fixed operators will be looking forward, understanding that a new technology is needed for further growth and looking for the next area of investment after their core network upgrades are complete. This will lead them to increasingly deploy fibre. Fibre to the home will be standard in new builds and an upgrade programme will progressively occur across developed countries. Necessarily this will be slow – for most countries a decade will be needed for a large percentage of the buried copper to be replaced. The timing of the build will vary dramatically from country to country. For countries like Japan that had already started by 2005, the build might mostly be complete by 2016. For countries like the USA, where the drivers for fibre are strong, the build might be half complete by 2016. For other countries like the UK, the build might start only around 2016, extending to 2026. So by 2016, a moderate number of subscribers across developed countries will have essentially unlimited data rates to the home. However, there will be few new services able to take advantage of these rates as yet – primarily the service provision will lag the deployment because it will be difficult to make a business case for the service while there are few potential target customers.

Core IP network deployment will be mostly complete across the developed world, allowing fixed networks to offer a wider range of end user services, and enabling all calls to be IP-based.

### Broadcast networks

By 2016, broadcasting will have started to change. There will still be a large percentage of the population that has the traditional TV in the living room that uses standard broadcast mechanisms to receive their signals. However, a major part of the population will have moved to a model whereby their PC 'assembles' their own TV channel for them, putting together a mix of broadcast material and downloaded material. Some of this will be packaged for podcasting, some for viewing, depending on preferences. For these individuals, the PC will have become the home media entertainment system, able to stream video to any display in the house as required using wireless networks. All viewers will be building significant video libraries on large hard disk storage systems which they can access in various ways throughout their home. High definition will be standard for a significant percentage of all TV channels.

Mobile TV networks will have been launched by this stage, and either failed to generate a commercial model, or proved to be successful. Which of these will transpire is still far from clear. Tentatively we are of the view that one network might succeed in each country offering news, sports and other programmes which have particular value in being live.

Digital switchover (from analogue broadcasting) will now be complete. The spectrum will be used predominantly for more terrestrial TV channels, including some local and some high-definition and limited mobile TV.

### Wide area wireless

We expect 3G to still be the system of choice by 2016. For many operators their 3G systems will be about 12 years old, comfortably within the lifetime of such a technology. With no

4G system on the horizon, the future will be 3G upgrades. A wide range of enhancements will have been implemented to the networks, increasing data rates and functionality.

Limited mesh network deployments might also have taken place. These will have a range of functionality. They will extend the range of cellular networks into buildings and other hard-to-cover areas. They might also extend W-LAN networks, allowing suburban coverage but at lower data rates and longer latency.

Between around 2010 and 2016 we expect a battle for the indoor environment to have taken place, which 3G will broadly have lost. Despite many designing small, inexpensive 3G home base stations, the ubiquity and usefulness of the W-LAN will have resulted in most in-building communications being carried on dedicated W-LAN networks. There will, however, remain a subset of users, such as those renting for short periods, who do not wish to have a fixed line into their home and instead rely entirely on 3G to provide them all their service. Broadly, cellular operators will have stopped attempting to be a key player in provision of services into the home and instead will concentrate their coverage efforts on excellent wide area service, including underground tunnels, train lines and rural areas.

Users will now be completely familiar with using their mobile for Internet browsing while on the move. Mostly this will be for contextual information such as weather, directions or nearby attractions. This will have become a way of life, and an inconvenience when not available. Video calling will be more prevalent, and socially acceptable for particular types of calls. It will never fully replace voice-only calls, but as network capacity grows and costs come down will gradually become the call type of preference.

### Short-range wireless
Air interfaces will see little change between 2011 and 2016. There will be some further evolution of W-LAN and BlueTooth but broadly data rates and functionality will be acceptable to all. Enhancements will be mainly focused in improving plug-and-play capabilities and overall device intelligence as well as working in high-interference environments.

Outdoors, hotspot growth will have levelled off, with all significant public places having hotspot coverage. Our assessment is that the initiatives to cover whole cities with W-LAN systems will have failed commercially. Users will have found that the combination of 3G, with its falling costs and improving data rates, and nomadic W-LAN coverage when static provides the best combination of cost and utility. However, it may be that the emergence of mesh technology changes this assessment somewhat by allowing lower cost deployment of mesh-networked W-LAN nodes outside of the hotspots. These will provide lower data rates and slower connectivity but this may be sufficient for many applications.

Indoors, W-LAN functionality will now be the norm, with the W-LAN providing telecoms support, transfer of video material and a wide range of other tasks around the home, and handling voice traffic in the office. W-LANs will use mesh or repeater functionality to ensure good coverage is achieved. Many home electrical appliances will now come with W-LAN capabilities that will enable security systems, major appliances, etc., to be controlled remotely. Early applications will be simple – such as alerts if the refrigerator goes outside of its temperature limits and the ability to optimise energy efficiency by using a central electricity control module that tracks changing energy prices. Because W-LAN chipsets will be so inexpensive, even if only a few users make some use of this functionality it is commercially viable to add it.

Widespread BlueTooth functionality will also have resulted in wireless connectivity emerging into unlikely devices. For example, running shoes may include functions to measure distance run, effort exerted etc., which can be transferred to a PC via BlueTooth.

Short-range wireless will be one of the predominant mechanisms used to update storage devices on handsets. As a handset moves into the home, office or other place where a high-speed short-range wireless system is available it will initiate an update process. This might update emails and attachments, music files and download video content that the user might be expected to watch, such as the latest news bulletin. By this means, users will often find that the video material they want is already on their handsets and will only need to use wide area networks when they are away from hotspots and want to access live material, such as a sporting event.

The user will now live in an environment where he or she expects that any electrical device will come with some wireless functionality. Simply introducing the device into the home and initiating a 'new device' function will result in the home network recognising the device and providing appropriate user functions – much as happens today when plugging a new USB device into a Windows PC. The idea of a home control hub will be growing. While this is based on the home PC, it may be accessed via a W-LAN tablet left around the house, or remotely using any form of connectivity.

### Handsets

The trends to add more functionality to the handset will increase. Cameras, music players and video player functionality will be standard. Internet capabilities will improve. Functions such as heart rate monitoring, GPS, exercise assistance, etc., will be available. Displays will be somewhat improved – brighter and clearer but no breakthroughs will occur in rollable or foldable displays. Most phones will make use of speech recognition, with network-based recognition widely used both to 'work' the phone and also to dictate emails and similar. Embedded recognition will be available as a fallback for when network connections are not possibly. Phones will all be multi-modal covering 2G, 3G, BlueTooth, W-LAN and possible UWB and RFID-type standards such as near-field communications (NFC). Phones will have large amounts of storage, often in excess of 100 GBytes, allowing a substantial amount of video and music content to be stored on the handset.

### Services

The key evolution during this period will be broadcast services, as discussed above. Services will also grow on the back of simplification, with the handset and network becoming increasingly able to provide the user with context-sensitive and personalised information and services.

### Convergence

The voice convergence will now be complete. Users will have a single phone they use in all environments. Filtering systems will present them with work and personal calls according to their preferences, location, time and other parameters. The broadcast convergence will now be well under way, with broadcast material increasingly accessed across telecommunication networks, stored on PCs and watched on a range of devices. However, this is a convergence that will continue for the next decade due to the slow speed of change for some broadcasting viewers.

*Industry structure*

We commented on the 'tipping point' phenomena for industry structure in our predictions for 2011. It seems likely that, with voice convergence complete, the tipping point will occur at some point between 2011 and 2016, likely nearer the start of the period than the end. This will result in operators splitting themselves into 'pipe' companies and service providers. The service providers will then seek to put together converged offerings across a range of pipe providers. This fragmentation might also extend to broadcasters who split into 'multiplex providers', 'channel assemblers' and 'content sources'. This splitting of companies will then lead to a range of mergers and acquisitions, with big-brand players entering the service provision market.

*Other significant issues*

During this period we may start to see the early deployment of a range of sensor networks. These might be networks, for example, of sensors clustered around transportation networks, ensuring traffic flow is normal, in the garden, checking on the conditions of plants and lawn or providing security-related information around homes and offices. These will be enabled by the combination of low-cost sensors; battery power systems that can recharge themselves from solar power or other sources, and low-data-rate mesh wireless technology that can link all the sensors together. Such networks might effectively extend the home wireless system, providing input data that can be used to control, for example, a sprinkler system in the garden.

*End user experience – how will it have changed*

Change is more of a continuum than a specific event. Building on the change of the previous five years, users will now become completely comfortable with convergence. The home phone will be relegated to a cupboard 'just in case'. Users will understand how to structure their communications in such an environment and when to use video calls. Users will perceive a significant revolution in broadcasting – no longer will many use TV guides. Users will also expect wireless control of all electrical devices in the home and office, including the ability to remotely interrogate home appliances from devices such as the cellphone.

---

**A day in the life – 2016**

It is Wednesday morning in the Potsman household. First to awake is John. His mobile acts as his alarm clock although it still has to be set manually. It randomly selects a track of music from his library to play to John who likes this little bit of variety to start the day. Today it was Wagner – rather too over the top for 6 am, he thought. He got up and had breakfast. While he was munching on his multi-vitamin cereal a flashing icon on his phone caught his eye – it was getting its daily download from the PVR. This was programmed to deliver a podcast to each member of the house at around 6.30 am.

John jumped on his bike to cycle to the station – it was good exercise and with the high price of fuel made good sense too. And, with his watch acting as a heart rate monitor and feeding information via BlueTooth to his phone, he could show he'd burnt off more calories this way – he might be allowed dessert later! When he got to the station he checked the departures board – no reported delays for a change. He'd have known if there were major disruptions anyway, he'd registered on the railway operator's website to receive emails if his preferred train was predicted to be more than 10 minutes overdue. Meanwhile, at home, the rest of the family was going through their morning routines.

On the train, John stated to watch his podcasts. His preferences were for a summary of the key financial news overnight, an update on key news around the world and then, if available, highlights of any cycling races that had taken place the previous day – John enjoyed cycling but didn't have time to watch the full race coverage each day. When these were finished, John switched to CNN Mobile to get a specially tailored version of the mobile news, broadcast over a special mobile TV network. This repeated every five minutes, so it didn't really matter when you started watching it.

John left his train and walked to the underground. While waiting on the station his phone buzzed. John pulled it out of the belt clip – it was an urgent message from the home security system, the front door had been locked without the security system being engaged. John sighed, his daughters often forgot all about security – perhaps it was because they grew up in an era when burglaries were very rare. John pulled up the browser, selected his homepage bookmark, clicked on the 'manage home' icon and turned on the security system remotely. He hoped nobody was still left in the house! If they were he'd soon know because he'd get another alert telling him the alarm had gone off. John was pleased that the underground now had cellphone coverage – it had taken ages for that to be installed, mostly because the operators couldn't agree terms with the underground company.

Later that day, John's wife Anna arrived back home. As was her habit she picked up the web tablet and clicked on location to see where all her family was. Family location was a service provided by the cellular operator to families where all members agreed. It was sometimes only accurate to a few hundred metres, but sufficient for Anna to check that everyone was where they should be. It looked like she had an hour to herself before the children got home. She sat down in the lounge with a cup of coffee, picked up her phone, which also worked as a remote control for the entertainment system and selected her section on the PVR. A list of the programmes recently recorded that she was most likely to find interesting appeared. She selected a travel show and relaxed for a while.

After that it was a normal hectic evening, with the girls to sort out, a meal to prepare and the washing to do. Later that evening John got home. 'It's been a long day', he said, 'why don't we pick a film, open a bottle of wine and relax together?' They turned on the widescreen display in the lounge and selected 'video on demand'. The entertainment system quickly connected to the Internet, filtered the films they had already seen and presented them with a list of those most likely to be of interest. They agreed on a recent thriller, accepted the payment terms and the system began the download. Because the download speed was faster than the viewing speed, the film started instantly and had completely streamed to their hard disk before they were half way through watching it.

## 15.4  Predictions for 2021

*Fixed networks*

By 2021, much of the fibre deployment around the developed world will have completed. Homes and offices will essentially have unlimited bandwidth with little charge for usage. With all traffic being carried over IP (updated to a new version) there will be complete integration of voice, video and data, allowing great flexibility in the ability to change call type during a call, multiplex multiple streams and for both the operator and the user to build services. Broadly, this will complete the evolution for fixed networks. Beyond increasing reliability, there will be no further need for investment for the foreseeable future.

*Broadcast networks*

The broadcasting 'revolution' will have predominantly happened by 2021. Almost all homes will use entertainment centres linked into the Internet as well as satellite broadcasting networks. In many countries terrestrial broadcasting will have been discontinued – the radio spectrum will have been judged too valuable to dedicate to a service that everyone could equally well access over the Internet. This will have annoyed many who will remember that it was only ten years ago that they were forced to buy a 'digi-box' to upgrade from analogue to digital TV reception. The concept of TV 'channels' will have also disappeared. Instead, producers such as the BBC will offer a range of content and PVRs will seek out and assemble 'personal channels' from the totality of all the content.

Mobile TV networks will continue to be useful, however, to provide updates for those on the move. Handsets will tune into these networks in idle periods, and store any information that appears interesting, so the viewer can see, for example, the latest news whenever he or she wishes.

*Wide area wireless*

Wireless networks will hardly change. Over time, cell sizes will become progressively smaller, capacity will continue to grow and data rates will expand. But with users accessing a variety of short-range networks for their high- data rate needs, dramatic advances in cellular systems will be seen as unnecessary, likely to create more unwanted masts and bring few benefits.

*Short-range wireless*

Short-range networks will also have stopped evolving significantly. They will be able to provide more than adequate data rates for all the applications in the home, office and hotspot, have excellent coverage and be extremely user friendly. Because of the large number of electrical devices that interact with them, any changes will need to be backwards compatible and relatively slow to occur.

Wireless control will increase. For example in all new homes it will be possible for all electrical functions to be remotely controlled, including lighting and heating, although there will still be wall switches for most occasions. Sensor networks will also be more widely deployed, so that homes and offices will be provided with a wide range of information about temperature, humidity and other data such as moisture levels to allow leaks to be detected quickly.

*Handsets*

Little will have changed over the last five years. Handsets will now have all the functionality and multi-modality that people require. Speech recognition, better software, some contextual awareness in the handset and familiarity will have made interaction with the handset much simpler, allowing users to employ the wide range of functionality with ease. However, battery life will remain an issue, preventing high-power displays or intensive processing within the phone.

### Services
The key area of growth will be in personalised services, able to perform tasks like setting a user's alarm clock, checking traffic forecasts appropriate to travel plans and so on. These will be network-based, sending messages to handsets or homes as needed.

### Convergence
By 2021, convergence will be complete, and little discussed. People will look back on an era where different services were provided by different operators via different channels with some amazement. Communications will clearly be divided into a range of pipes, a range of services and a selection of content. Any service or content will be available via any pipe, with the best selected depending on location, requirements and preferences.

### Industry structure
The industry structure will now be completely aligned with the converged services. There will be network operators, service providers and content providers. Users will select their preferred service provider, who they believe will best meet their requirements and who they trust.

### End user experience – how will it have changed?
Users will become increasingly familiar with their new converged world. They will perceive increased value in personalised services, allowing their service provider, via their phone, to make all sorts of decisions and provide suggestions for them.

## 15.5 Predictions for 2026

We expect little further change between 2021 and 2026. In essence, fixed networks, wide area networks and in-building networks will all be able to meet all the requirements of the end user. Convergence will be complete. Levels of personalisation will increase but the capabilities of the handset will level out, unless an unexpected breakthrough in display technology occurs. Just like the railways in 2005, communications networks will have become stable with little further change predicted. As far as users are concerned they will be more than adequate – so much so that their presence is completely taken for granted and little thought about.

---

**A day in the life – 2026**

Ten years later in the Potsman household and life has changed a little. The girls are now at universities and Anna has returned to work. John, however, is still commuting into London. When his mobile alarm goes off this morning he awakes with a start,

*(continued)*

not just because it has picked Beethoven's Fifth to awake him with, but because it is still dark – why has his mobile woken him so early? He picks up his phone. The information he wants is already on the display – the phone knows that this is the first thing he will want to see. It tells him that there has been a signalling failure on the train network near Cambridge. It recommends he drives to a different station which is unaffected by the problem before catching the train. John sighs, no cycling to the station today then. On the plus side, the phone had sent a command to the central heating system to come on earlier, so at least it is warm and there is plenty of hot water.

As John gets into his car to leave for work, the mobile links via BlueTooth into the car audio system. John's mobile has Galileo navigation and relays voice commands providing driving directions. At the station John has no need for a modified train ticket. The mobile had earlier purchased the necessary ticket on-line and stored the confirmation code in its memory. On passing through the barrier at the station John places his phone near the reader and the phone provides the confirmation code using a short-range 'near-field' communication system. On the train, as before John views his selection of podcasts. Over the years, this service has gradually improved, with the PVR able not only to decide which pieces of content John will be most interested in, but also to edit them to remove superfluous material. It stores additional material, providing links if John finds a particular item to be of interest. Today, an item on a comprehensive bio-monitoring system for racing cyclists captures his interest. He is too old now to race bikes but he still retains a keen interest in the sport.

When he gets to work, he no longer has any need to turn his phone off in meetings. He subscribes to a comprehensive message filtering system. This links to his diary, his location information and a detailed table of preferences he spent one Saturday establishing. The system considers who each incoming call is from, any indication of urgency provided and the format of the call. It then looks at John's whereabouts and decides whether to deliver the call, to send it to a mailbox or to translate it into a different format. John knows that when in a meeting his phone will never make a noise. It might vibrate discretely if there is an important message for him.

Anna is also at work. After a busy morning she is sitting in the local coffee shop wondering what they should have for dinner that evening. She has found it very useful to subscribe to the 'Good Food' meal planning service. Each day this sends her suggestions for the evening meal based on her preferences, previous selections she has made and the likely contents of her refrigerator. The latter is not perfect – it works by uploading the bill from her weekly grocery shop and then removing those items it deduces she has used for meals earlier in the week or which are out of date. Unfortunately, it doesn't always know about John's habit of snacking on whatever he finds in the fridge! She decides on a pasta dish for that night and selects the recipe. The web link will be automatically forwarded to the home network, so that it can be quickly recalled on the web tablet in the kitchen. She then turns to have a look at her podcasts while finishing lunch.

During the day, John doesn't bother to check on the status of the trains. His mobile will do that for him, it will also remember that his car is at a different station. He doesn't even need to bother much about battery life any more. All the desks at work have inductive charging pads inserted. Just lying the mobile, or any other battery powered device, on top of these recharges it. John also has one in his car and they have just installed them at home, on their bedside tables so that their devices charge overnight. In any case, if the mobile detects a low battery it will intelligently shut down elements of functionality and warn John that recharging will be necessary soon.

On the way home the trains are working well, but there has been a burst water main on one of the main roads near home. The phone is monitoring traffic information and automatically routes him around the affected area. It recalculates his arrival time home and automatically sends a message to Anna informing her of a slight delay – another service offered by their service provider. Anna was just about to start cooking – she goes back to browsing the web for a few more minutes.

After dinner, their home entertainment system reminds them that they have an appointment to call Lucy, their eldest daughter tonight. They routinely use video calling for this, sitting on the sofa and using the main flat panel display in the lounge. Lucy is in her bedroom at her Halls of Residence, all of which are routinely equipped with video calling capabilities. Anna, particularly, likes video calling – it makes it so much more personal to be able to see each other. She admitted that she was initially sceptical about whether she needed video calling – but like so many aspects of her electronically organised life she wouldn't go without it now.

As they go to bed that night, John idly wonders at what time and with what piece of music his phone will wake him tomorrow...

## 15.6  Key New Applications

It is applications that drive user acceptance of new technology. In the sections above we have described a large range of applications that might be used in the next 20 years. To attempt to describe all the applications that might emerge would be near-impossible – it would be akin to trying to predict all the different types of software applications that might be developed for a PC. However, it is possible to discuss more generally the different classes of application that we expect to see, and that have been illustrated above. These include:

- an ability to retrieve and listen or view audio/visual information, such as TV programming, regardless of location or device type
- an ability to monitor the environment, such as the temperature of each room, the status of devices, the moisture levels in the garden, etc.
- an ability to stay connected with high-speed data connections in most urban environments
- applications to control the home environment and provide audio-visual content and communications in the home environment

- applications to interwork personal devices, allowing them to exchange information, synchronise and make use of nearby facilities such as larger screens
- applications based on understanding location
- an ability to make video calls whenever required
- applications based around transport including routing, traffic avoidance and ticketing
- a wide range of personal applications which use wired and wireless networks as underlying bearers such as providing detailed personal organisation, suggesting meals based on the contents of the fridge, etc.

## 15.7 Key New Technologies

The new technologies that we expect to develop and be used to deliver some of these key applications are detailed in Table 15.1.

**Table 15.1**  Key new technologies.

| Application | Technology/spectrum/ regulation | Other advances |
|---|---|---|
| Retrieve and watch audio/visual information | Additional spectrum allocation perhaps at UHF | Larger storage on handsets to allow podcasting, intelligent PVRs to push content to the handset. |
| Monitoring of environment | Mesh networks, unlicensed spectrum, possibly lower cost RFIDs | Better batteries, cost reduction for sensors |
| High-speed connections in urban environments | Possibly mesh backhaul to extend W-LAN coverage, additional spectrum | |
| Home networks | Cheaper W-LAN chips to allow them to be more widely embedded, possibly UWB | Better software to manage home networks, W-LAN integrated into major appliances |
| Personal networks | Possibly UWB | Better software, more wireless enabled devices |
| Location services | | GPS deployed in handsets |
| Video calling | Possibly more spectrum | |
| Transport guidance | Sensor networks (see monitoring environment, above), location | Integrated transport information, intelligent handsets |

Many of the applications envisaged will make use of a mix of these technologies. For example, the ability of the handset to wake the user earlier in the morning when there are traffic problems might require 'transport guidance', 'home networks' and possibly 'personal networks'. As mentioned in earlier chapters, there are some technologies that we do not see as significant (although some contributors disagreed) and are missing from this table including SDR, cognitive radio and MIMO antenna systems.

## 15.8 Key Changes in Networks

Table 15.2 details predicted changes in wireless networks.

**Table 15.2**  Changes in networks.

| Network | Key changes | Driver for change | Enablers required |
|---|---|---|---|
| Cellular | Coverage extended e.g. into tunnels, capacity enhanced through more cells | Slowly increasing capacity demand as applications like video calling take off | None, although lower cost backhaul would aid economics |
| Public W-LAN | Coverage extended both directly through new installations and indirectly through mesh linkage | Increased demand for downloads across major urban areas | Advances in mesh technology, as above lower cost backhaul also valuable |
| Broadband fixed wireless access | None – we do not expect to see widespread successful commercial deployment | N/A | N/A |
| Office and home | W-LAN increasingly ubiquitous | W-LAN embedded into more devices | Lower cost W-LAN chipsets with better battery life, possibly better interference mitigation technology |
| Satellite | Minimal | None – satellites still needed for coverage of remote areas, seas etc. | N/A |
| Emergency services | Possible enhancements for niche areas such as coverage of burning buildings with detailed location information | More efficient and effective ways of working | Mesh technology, advances in device processing power |
| Transport-related | A range of new specialised networks emerge to monitor traffic conditions and related data | Demand for better traffic guidance | See 'Transport guidance' in Table 15.1 |

## 15.9  Major Growth Areas

Based on the analysis presented above, the areas we expect to see the most significant growth in terms of sales, value created by companies, etc, are as follows.

- *Handsets.* These will become more advanced, with greater processing power, more integration of other devices, advanced screens, etc. As a result they will be higher price, sold in large volumes and changed frequently. This will drive growth in sub-elements, including advanced screens, better batteries and small high-density storage devices.
- *Home networks.* Whether these evolve from PCs or from current modems, they will require substantial memory and processing power and multiple means of interconnection. They will also require complex software able to handle the widest range of different home configurations.
- *Contextually aware software.* In order to provide the intelligence to allow handsets to predict user requirements, advanced software able to understand a wide range of contextual inputs will be needed.
- *Network software and servers providing convergence.* Complex software will be needed in the network to ensure that whatever communications channels and devices are available; the user is kept connected in the best manner possible.
- *User applications.* In the same manner that much value has recently been created by applications running on top of Internet infrastructure such as Google, eBay and MySpace, we expect there to be many valuable applications written for wireless networks.
- *Service provisioning.* There will be increasing value in a service provider that enables multiple different communications channels, provides an 'IT' fault diagnosis service, screens for junk messages and viruses and generally keeps a user's communications environment 'working'.

Where would we invest our own money? This changes over time. During the period 2006–2011 the key focus is on deploying the underlying wireless capabilities. Companies that can deliver advanced home networks and context-aware software for handsets and networks will do particularly well. After around 2011 the focus will switch to applications and we would invest in companies developing innovative applications for mobile devices, of which we expect there to be a very broad range.

## 15.10  Areas we Predict Will not be Successful

There are a number of technologies and services under development today that we have predicted will not be successful (although the contributors tended to disagree with many of these).

- *SDR.* We believe that multi-modal phones will be cheaper and just as effective.
- *Cognitive radio.* We note a number of difficulties and cannot see any convincing application.
- *Smart or MIMO antennas.* As cells get smaller, the benefits of these fall while the cost per user increases.

- *Fixed wireless access.* We believe wired technologies will continue to be cheaper and offer higher data rates in all but a few niche applications.
- *4G.* We do not see the need for a completely new generation, nor the 'space' where it will provide distinct advantages.

## 15.11  Implications for Stakeholders

In this section, we consider what the changes predicted here might mean for different groups of stakeholders.

### Manufacturers

For manufacturers, this forecast presents a picture of two halves. In the early years there will be considerable deployment of new technology, including 3G networks, mobile TV networks and core IP networks. Sales of handsets, W-LAN routers and BlueTooth devices will grow considerably. However, between 2011 and 2016 networks will become stable and little additional roll-out will occur. Of course, there will continue to be upgrades, replacement of faulty and obsolete equipment and filling in additional cell sites, but the days of complete network builds will be over.

Handset manufacturers are likely to see continued growth, with handsets developing ever increased capabilities and better human interfaces. There will need to be on-going research across a wide range of areas to improve capabilities such as displays, batteries and speech recognition.

The immediate steps that manufacturers could embark upon are:

- Gradually move R&D effort from networks towards handsets.
- Increase pressure on infrastructure divisions to downsize, or at least maintain size.
- Enhance research efforts in areas of convergence.

### Operators

For operators too, this is a future with two distinct periods. In the first decade or thereabouts, for most operators this will be a period of stability and profitability. No major change in plans will be needed from any operators. Perhaps operators might use this time to focus on reducing costs, improving networks and building portfolios of new services. Then, at some point, the industry structure will shift. Operators will undergo major and painful splits with a subsequent period of mergers and acquisitions. This will distract the operators for some time, but once over this period there will be a strong growth in service offerings, potentially requiring enhancements to core networks.

The immediate steps for operators are to:

- Draw up plans to utilise the period of stability to optimise operations.
- Consider outsourcing network build and operations as a precursor to a split into pipe and service organisations.
- Actively pursue collaborations for delivering converged services.
- Draw up 'war plans' as to strategies to adopt when the industry split occurs.

**Service Providers**

Pure service providers will continue to find it difficult to operate profitably until the industry structure changes as described above. At this point there will be a host of service providers created from existing operators. Many new ones will also likely enter this new market. There is probably little that the existing service providers can do in the interim.

**Regulators**

There are surprisingly few implications for regulators. Concerning spectrum regulation there is little additional demand predicted for spectrum in our view of the future. Pressure will likely increase on unlicensed spectrum and there may be opportunities to reduce this pressure through a combination of additional spectrum and 'rules' such as politeness protocols, which have been proven to increase capacity. Regulators could start to research and consult on their approach to unlicensed spectrum now. Concerning competition regulation, little will change from today's position. It may be that some partnerships to provide convergence are judged anti-competitive. However, after the change in industry structure, there should be increased competition and hence less need for regulation. Hence, regulators might consider now how they could reduce regulation as this future unfolds.

**Academics and Researchers**

Our predictions suggest that the key areas where advances will be needed are:

- handset technology including batteries, displays, man–machine interface, processing power and storage
- software capable of providing contextual information and eventually leading to automated diary management and environment control
- systems that can handle the complexity of a large number and wide range of wireless devices (e.g. in the home), but present a simple plug-and-play interface to the user
- backhaul systems that will facilitate the rapid and inexpensive deployment of cells.

In addition, although not discussed here, there are likely to be continual requirements for improving security. There may also be a need to progressively reduce personal exposure to RF emissions.

This agenda is somewhat different from the current research profile. At present, there is still significant effort expended on faster air interfaces or means to provide more efficient throughput. This includes research into MIMO and smart antennas, complex OFDM approaches and mesh networks. Our predictions suggest that much of this work is unnecessary. There is also much work on enhancing the interaction between different layers in a system to optimise across multiple layers. This may be more appropriate as a low-cost mechanism to gain capacity.

Much of what is currently being researched could be termed the 'traditional' topic areas. In a world where there is unlikely to be a completely new standard, a different approach is needed. This will likely be difficult for academia, where key staff may wish to continue directing research in their own areas of expertise. However, much research is industry funded, so it is likely that academia will eventually follow an industry lead towards the topics listed above.

## 15.12 Differences from the Prediction Made in 2000

Table 15.3 below sets out a range of predictions made in 2000 and in this book and provides a direct comparison. We then discuss the trends and implications of this in more detail below.

Reformatting the table around relative timings leads to Table 15.4.

The reasons for the major differences in prediction are as follows:

- *Filtering and redirection.* It was assumed that voice convergence would create a need for this service. With our prediction that voice convergence may now be anywhere between one and six years later than previously forecast, filtering and redirection also are deferred. Voice convergence has been delayed because the industry structure is not appropriate to promote it, with few operators having both a mobile and a fixed presence. Many mobile operators have been pursuing a 'mobile only' approach. The impact of industry structure on this prediction is one of the reasons we have considered industry structure in more detail in this forecast.
- *Core IP networks.* This has proved to be a larger task than expected. Standards have not developed as quickly as foreseen and operators have proven to be risk averse in moving from their existing working systems to new unproven ones. The lesson here is that large-scale infrastructure change is likely to be a very slow process.
- *Car linked to home.* This was an error in the original forecast. Car manufacturers are typically very slow to integrate new technology into the car, because with a lifetime of ten years or more, if the wrong technology is selected it can generate long-term problems or dissatisfaction. Hence the reason why BlueTooth is only now being widely deployed. When considering adding telecommunications electronics to cars it needs to be remembered that the automotive industry will likely wait around five years after the successful introduction of a new technology before it considers it sufficiently stable to put into mass production vehicles.
- *Navigation standard.* Adding GPS to handsets has proven somewhat more expensive than originally forecast. Also, the demand for location-based services to date has been weaker than expected. The lesson here is that introducing new services takes some considerable time if users need to change their behaviour before the service becomes valuable.
- *Handset managing environment.* Two errors were made in the original prediction. The first was an assumption that other devices, such a home thermostats, would rapidly gain wireless functionality. While this is technically simple, the drivers to encourage users to pay for such an upgrade have proven weak. The second was an underestimation of the complexity of a device able to handle many different devices in a friendly, intuitive manner and control them appropriately. As has been shown on PCs, it has taken many versions of Windows until even the connection of devices to a computer has become straightforward.
- *Advanced display mechanisms such as retinal projection.* These have proved more difficult than expected. Advances in displays have been slower than anticipated and the slow advance in battery life has proved a significant problem for novel displays which tend to be higher power. There is now significant doubt as to whether any major progress will be made in this area in the next 20 years.

**Table 15.3**   Differences between predictions of 2000 and 2006.

| Prediction | Date predicted in 2000 | Date predicted in 2006 | How the 2006 prediction compares |
|---|---|---|---|
| Homes deploy wireless networks | 2005–2010 | 2006–2011 | Aligned |
| Handsets work on home networks | 2005–2010 | 2011 | Slightly later |
| Filtering and redirection widespread | 2005–2010 | 2016 | 5 years later |
| Video calls available but rarely used | 2005–2010 | 2011 | Slightly later |
| Broadband to home at 10 Mbits/s | 2010 | 2011 (via ADSL 2+) | Aligned |
| Speech recognition standard | 2010 | 2011 (network based) | Aligned |
| Core IP networks commonplace | 2010 | 2016 | 5 years later |
| Public W-LAN hotspots ubiquitous | 2010 | 2011 | Aligned |
| Car linked to home network and mobile | 2010 | 2011–2016 | 5 years later |
| Navigation functions standard and intelligent | 2010 | 2011–2016 | 5 years later |
| Home wireless networks and devices ubiquitous | 2010–2015 | 2011–2016 | Aligned |
| Video calls reach 50% of all calls | 2010–2015 | 2016 | Slightly later |
| Broadband to home at very high data rates | 2010–2015 | 2011–2021 | Slightly later |
| Core IP networks ubiquitous | 2015 | 2016–2021 | 5 years later |
| Fourth-generation cellular deployed | 2015 | (not anticipated) | |
| All machines in homes have wireless communications | 2015–2020 | 2016–2021 | Aligned |
| Handset intelligently sets daily agenda | 2020 | 2016–2021 | Aligned |
| Handset 'manages' home, car and local environment (eg hotel room) | 2020 | 2021–2026 | 5 years later |
| Handset able to use nearby screens | 2020 | 2021 | Aligned |
| e-payment | 2010–2020 | 2011 | Aligned – but via a different mechanism |
| Retinal projection or other novel form factors | 2010–2020 | 2021 or later | At least 5 years later |
| Mobile TV emerging | (not predicted) | 2011–2016 | |
| PVRs widespread | (not discussed) | 2011 | |

**Table 15.4**  Areas where forecasts aligned and differed.

| Aligned on | Slightly later on | 5 years later on |
| --- | --- | --- |
| Home wireless deployment and devices | Handsets working in the home | Filtering and redirection |
| 10 Mbits/s Broadband to home | Video calls | Core IP networks deployed |
| Public W-LAN | 100 Mbits/s broadband to home | Car linked to home |
| Handset setting daily agenda and payment |  | Navigation functions standard |
| Speech recognition |  | Handset managing environment |
| Ability to use nearby screens |  | Retinal projection and similar |

Overall, a clear message is that all the previous predictions either still appear likely or were optimistic – none were pessimistic. This suggests a tendency towards assuming that more will happen over the next 10–20 years than will actually be the case. We have tried to correct for this tendency in our new forecast (although there is always the risk of swinging too far the other way!). Disregarding the timings, the predictions are still broadly the same in terms of the capabilities of technologies, the data rates achieved and the services that will be provided to the end users. Given that little has changed in the communications environment in the last five to six years, perhaps this should not be too much of a surprise.

## 15.13  The Future in a Page

We believe that it is possible to predict the next 20 years of wireless with reasonable certainty. For the user, the next 20 years will see a very substantial, but steady change. Users will come to rely on their handset as a single device to manage their communications and, indeed, much of their life. It will truly become a 'remote control on life', with massively enhanced capabilities including huge storage, advanced methods of user interaction such as speech recognition and many in-built tools such as cameras, music players etc. Users will cease to differentiate between different communications channels and instead see the world as one large communications network, able to provide them whatever content they need wherever they are. Users will also no longer see broadcasting and communications as separate; indeed, the concept of broadcasting will change dramatically to one of content provision that is sought out by users – more like the publishing model of today. Users will perceive their lives becoming more convenient both in the home and wider area. At home, their home wireless systems will automate a range of tasks and provide new capabilities such as suggesting menus based on the contents of the refrigerator. Out of the home, their devices will book and alter travel according to conditions, manage diaries and ensure appropriate information is available.

Achieving all of this will require little in the way of change for wireless technology, which is already capable of delivering more than adequate data rates and services if deployed with sufficient density – and indeed no further significant advances in wireless technology are expected. As a result, no new generations of wireless network or widespread network deployments are predicted, although existing networks will be much enhanced. However, there will be substantial progress in the intelligent systems that use context to configure devices appropriately, control interaction with the handset, and control home and office networks in a simple yet intelligent manner. Battery and backhaul will remain areas where substantial progress would make a significant difference, but the barriers will be such that only steady improvements can be expected.

While technology may change little, achieving the vision of the future set out here is dependent upon changing the current industry structure from one of vertically integrated providers, owning both the network and service provision to a world where there are network owners, content providers and service providers. We expect this change to occur quite suddenly at some point in the next decade. Change also requires consumer willingness to adopt, which history has shown is often a slow and unreliable process.

Overall, the future is marked by an initial period of stability as 3G and broadband networks are built out followed by a short period of upheaval as the industry structure changes dramatically and new services and service providers emerge – this is the point at which convergence truly happens. Beyond this we expect the underlying wireless communications infrastructure to become a slow-changing utility similar, for example, to railways or increasingly the core Internet infrastructure, but with substantial excitement and growth around the services provided on top of this wireless platform.

## 15.14 . . . And the Elevator Pitch

The mobile phone will increasingly become an indispensable tool to navigate through the day, truly acting as a 'remote control on life'. Communications channels will converge such that any content is available through any channel. Little new technology will be needed to realise this, but the industry structure will undergo major change with new service providers emerging.

# List of Acronyms

| | |
|---|---|
| 3GPP | 3G Partnership Program |
| ADSL | Asymmetric Digital Subscriber Line |
| AEHF | Advanced EHF |
| ALE | Automatic Link Establishment |
| AM | Amplitude Modulation |
| AP | Access Point |
| ARPU | Average Revenue Per User |
| ARQ | Automatic Request Repeat |
| ATM | Asynchronous Transfer Mode |
| AVL | Automatic Vehicle Location |
| BER | Bit Error Rate |
| BLOS | Beyond Line Of Sight |
| CCFL | Cold Cathode Fluorescent Lamps |
| CDMA | Code Division Multiple Access |
| CDMA 1X | CDMA using 1.25 MHz bandwidth channels (one times or the same as existing channels) |
| CDMA 1X DO | CDMA 1X Data Only |
| CDMA 1X DV | CDMA 1X Data and Voice |
| CEPT | Conférence Européenne des Administrations des Postes et des Télécommunications |
| CLASS | Customized Local Access Supplementary Services |
| COMSEC | Communications Security |
| COTS | Commercial Off The Shelf |
| CR | Cognitive Radio |
| CSAC | Chip-Scale Atomic Clock |
| CSMA/CA | Carrier Sense Multiple Access with Collision Avoidance |
| DECT | Digital Enhanced Cordless Telephone |
| DMB | Digital Multimedia Broadcasting |
| DRM | Digital Radio Mondial |
| DSL | Digital Subscriber Line |
| DSP | Digital Signal Processor |

*Wireless Communications: The Future*   William Webb
© 2007 John Wiley & Sons, Ltd

| | |
|---|---|
| DTN | Disruption Tolerant Networks |
| DVB-H | Digital Video Broadcasting – to Handheld devices |
| DVD | Digital Versatile Disk |
| DWDM | Dense Wave Division Multiplexing |
| EDGE | Enhanced Data Rates for Global Evolution |
| EHF | Extremely High Frequency |
| EMP | Electro-Magnetic Pulse |
| EPOS | Electronic Point-Of-Sale |
| ETSI | European Telecommunications Standards Institute |
| EW | Electronic Warfare |
| FCC | Federal Communications Commission |
| FDD | Frequency Division Duplex |
| FDMA | Frequency Division Multiple Access |
| FM | Frequency Modulation |
| FSO | Free-Space Optics |
| FTTC | Fibre To The Curb |
| FTTH | Fibre To The Home |
| GEO | Geostationary (satellite) |
| GIG | Global Information Grid |
| GII | Global Information Infrastructure |
| GPRS | General Packet Radio Service |
| GPS | Global Positioning System |
| GSM | Global System for Mobile Communications |
| HAP | High Altitude Platform |
| HDTV | High-Definition Television |
| HF | High Frequency |
| HSCSD | High-Speed Circuit Switched Data |
| HSDPA | High-Speed Downlink Packet Access |
| HSPA | High-Speed Packet Access |
| HSUPA | High-Speed Uplink Packet Access |
| IC | Interference Cancellation |
| IEEE | Institute of Electrical and Electronic Engineers |
| IET | Institution of Engineering and Technology |
| IFRB | International Frequency Registration Board |
| IMS | IP Multimedia System |
| IMT 2000 | International Mobile Telecommunications for the year 2000 |
| IP | Internet Protocol |
| ISDN | Integrated Services Digital Network |
| ISI | Inter-symbol interference |
| ISM | Industrial, Scientific and Medical |
| ISP | Internet Service Provider |
| ISR | Intelligence, Surveillance and Reconnaissance |
| IPR | Intellectual Property Rights |
| ITU | International Telecommunications Union |

| | |
|---|---|
| JTRS | Joint Tactical Radio System |
| LAN | Local Area Network |
| LPI | Low Probability of Intercept |
| LPJ | Low Probability of Jamming |
| LTPS | Low-Temperature Polysilicon |
| LCD | Liquid Crystal Display |
| LO | Low Observability |
| LPE | Low Probability of Exploitation |
| MAN | Metropolitan Area Network |
| MIMO | Multiple Input Multiple Output |
| MoU | Minutes of Use |
| MoU | Memorandum of Understanding (when used in the context of operators) |
| MPEG | Motion Picture Experts Group |
| MMI | Man Machine Interface |
| MMS | Multi-media Messaging Service |
| MP-3 | Motion Picture Experts Group, Audio Layer 3 |
| MPEG-4 | Motion Picture Experts Group #4 |
| MPLS | Multi-Protocol Label Switching |
| MVNO | Mobile Virtual Network Operator |
| NBC | Nuclear, Biological and Chemical |
| NCO | Network Centric Operations |
| NEST | Naval EHF/SHF Terminal |
| NFC | Near-Field Communications |
| NIST | National Institute of Science and Technology |
| NVIS | Near-Vertical Incidence Skywave |
| OFDM | Orthogonal Frequency Division Multiplexing |
| OLED | Organic Light Emitting Diode |
| OTW | Other Than War |
| PAN | Personal Area Network |
| PC | Personal Computer |
| PDA | Personal Digital Assistant |
| PKI | Public Key Infrastructure |
| PMR | Private Mobile Radio |
| PRR | Personal Role Radios |
| PSTN | Public Switched Telephone Network |
| PTO | Post and Telecommunication Operator |
| PVR | Personal Video Recorder |
| QAM | Quadrature Amplitude Modulation |
| QoS | Quality of Service |
| QPSK | Quadrature Phase Shift Keying |
| R&D | Research and Development |
| RTS | Request To Send |
| RF | Radio Frequency |
| RFID | Radio Frequency Identification |
| SDR | Software-Defined Radio |
| SHF | Super High Frequency |

| | |
|---|---|
| SINR | Signal to Interference plus Noise Ratio |
| SIP | Session Initiation Protocol |
| SIR | Signal to Interference Ratio |
| SoHo | Small Office/Home Office |
| SME | Small and Medium Enterprise |
| SMS | Short Message Service |
| SNR | Signal to Noise Ratio |
| SSB | Single Side Band |
| TDD | Time Division Duplex |
| TDMA | Time Division Multiple Access |
| TD-SCDMA | Time Division – Synchronous CDMA |
| TETRA | Terrestrial Trunked Radio |
| TSAT | Transformational Satellite Communication System |
| TRANSEC | Transmission Security |
| UAV | Unmanned Air Vehicle |
| UFO | UHF Follow-On |
| UGV | Unmanned Ground Vehicle |
| UHF | Ultra High Frequency |
| UMA | Unlicensed Mobile Access |
| UMTS | Universal Mobile Telecommunications System |
| U-NII | Unlicensed National Information Infrastructure |
| USV | Unmanned Surface Vehicle |
| UUV | Unmanned Underwater Vehicle |
| UV | Unmanned Vehicle |
| UWB | Ultra Wide Band |
| VLF | Very Low Frequency |
| VoD | Video on Demand |
| VoIP | Voice over Internet Protocol |
| WAP | Wireless Application Protocol |
| W-CDMA | Wideband Code Division Multiple Access |
| WISP | Wireless ISP |
| W-LAN | Wireless Local Area Network |
| WLL | Wireless Local Loop |
| WMAN | Wireless MAN |
| WPAN | Wireless PAN |

# Index

*Wireless Communications: The Future*   William Webb
© 2007 John Wiley & Sons, Ltd

Printed and bound by CPI Group (UK) Ltd, Croydon, CR0 4YY

27/10/2024

14580151-0001